EXTINCTION
NOT THE END OF THE WORLD?

Steve Parker

Published by the Natural History Museum, London

First published by the Natural History Museum, Cromwell Road, London SW7 5BD
© Natural History Museum, London, 2013

The Author has asserted his right to be identified as the Author of this work
under the Copyright, Designs and Patents Act 1988

ISBN 978 0 565093211

A catalogue record for this book is available from the British Library

Designed by Mercer Design, London
Reproduction by Saxon Digital Services
Printed in China by C&C Offset Printing Co., Ltd.

Front cover: background image © PEKKA PARVIAINEN/Science Photo Library; p.21 *Smilodon*
and p. 73 California condor, © NHMPL.
Back cover p.12 *T. rex*, p.8 giant sable antelope, p.92 mammoth, p.41 elephant bird egg,
p.41 great auk, p.91 *Paranthropus*. All © NHMPL

contents

end and beginning
THE NATURE OF EXTINCTION

COIL-SHELLED AMMONITES, CRAWLING trilobites, massive-tusked woolly mammoths, fearsome sabre-tooth cats, the delicate Saint Helena olive tree, the dodo – all are famous partly because they are no longer with us. They have died out, gone for ever, exist no more. They are extinct. And they are far from alone. The vast majority of different kinds, or species, of living things that ever existed on Earth, are also extinct.

Looking back more than 3,000 million (3 billion) years, the early life-forms on Earth probably resembled micro-specks of jelly-like material in the oceans. The story of life tells us that they gradually evolved to become more diverse, bigger and more complex. Larger plants and animals appeared, first in the water, then invaded the land. This happened by the evolutionary process known as speciation. For example, imagine that the range of a particular kind, or species, of animal was split in two when an earthquake created a new river in the landscape. Isolated on either side of the river, each group of animals followed its own evolutionary path. As conditions changed, the animals did too. Eventually one species became two.

It is a common impression that by speciation, evolution has slowly fashioned greater numbers of more diverse species, up to the present day. It has gradually filled up the world to culminate in the amazing range of plants, creatures and other organisms that now populate the Earth with us. But in fact, life-forms became larger and more complicated quite rapidly. Earth's varied habitats soon had plenty of occupants. Since those times, hundreds of millions of years ago, there has been not so much an increase in species, but a turnover of species. For new ones to appear, others made way and disappeared – they became extinct.

So, extinctions are not unusual, recent or the exception. Indeed, they are the rule. They have happened, estimates suggest, to more than 999 of every 1,000 species that have ever lived on the planet. As part of evolution, extinction is not the end of the world. Demise of the old allows emergence of the new. It is integral to the Earth's continually changing range and richness of life-forms, known as biodiversity.

● **DEAD AS A...**
Not only dodos disappeared from Mauritius (see p.57). The island's dome-backed giant tortoise, and dozens of other animals and plants, have gone for ever.

Fossils and other evidence show that extinctions have happened regularly or in small, short bursts through most of prehistory. But here and there were 'major extinction events', often known as mass extinctions. During a relatively short period of geological time, sometimes much less than one million years, a great proportion of all species – one-third, one-half, even 90 per cent – disappeared.

Recent years have seen growing concerns that another mass extinction is upon us – and that its cause **is** us. It's often claimed that humans are altering the planet so much, and so quickly, that species are disappearing at record rates. Habitat destruction, pollution, global warming, over-exploitation, hunting, invasive plants and animals – the threats are numerous, massive and intertwined. However on the other side of the coin, speciation has not speeded up. With such rapid changes today, new species have no time to evolve and fill the gaps. So overall biodiversity is falling and the situation can only worsen.

How do these views stand up to scrutiny? And if they seem convincing, what can we do about the rash of extinctions happening now and likely in the future? Indeed, should we bother to do anything at all...

● DINOSAUR NO MORE
The great three-horned *Triceratops* is no more. But changing views on animal evolution and relationships mean that some dinosaurs still survive – see p.30!

local to global
TYPES OF EXTINCTION

PUT SIMPLY, 'EXTINCTION' MEANS that a particular kind of living thing has died out, no longer exists, and is gone for ever. Of course, very little in life sciences is that simple.

First, what living thing goes extinct? The term usually applies to a particular group of living things – most often a species. This is the basic, fundamental and natural unit in the biological classification of life-forms. A species is generally defined as a group of organisms capable of interbreeding and producing fertile offspring. (Different species may be able to mate, and even produce viable young, but those young cannot themselves reproduce and so are unviable.) Lately science has brought new facets to the definition of a species, such as the degree of variation in genetic material (DNA). But, for general purposes, the traditional description is helpful as the working yardstick.

In most cases, the members of a species look similar to each other. We have no trouble in recognizing them as belonging to their group. This is especially the case with the big, well-known animals popular with the public. So, among big cats, for example, lions are one species, tigers are another, and so on. Each of these species has a unique two-part scientific name that is internationally agreed and understood (and expressed in *italics*). The lion is *Panthera leo*, the tiger *Panthera tigris*.

An example of a recently extinct species is the huia, *Heteralocha acutirostris*. This was a type of wattlebird from the forests of North Island, New Zealand. Huia were probably quite rare before Europeans arrived. They disappeared during the early twentieth century – a victim of hunting for their highly valued feathers, and of deforestation to make way for agriculture. Huia were also sought after as mounted ('stuffed') specimens because the species showed a very marked difference in bills between the sexes. The male's beak was short, straight and robust, while the female's was longer, slim and down-curved. These types of dissimilarities between males and females of the same species are termed sexual dimorphism. Nowadays, they are well known in many different animals, from gorillas to butterflies. In the early twentieth century sexual dimorphism was less understood but a source of great wonderment. Thus the preserved female-and-male huia duo, with their oddly distinctive

● **HUIA**
Preserved female (left) and male (right) huia show sexual differences in bill shape, and so were highly sought after as 'curiosities of nature'.

● **HUIA FEATHER**
These feathers were much prized as adornments by Maori chiefs and then later collected by Europeans.

bills, became a much sought-after curiosity exhibit for collectors and museums. Today we would suggest that the different bill shapes evolved to allow females to exploit different foods to males, so the sexes did not compete. The female was able to probe into holes and crevices for insect food, while the male dug grubs and bugs from rotten wood. However we may well never know the details, since there were few detailed observations on living huia before their extinction.

Not just species

The huia is a clear-cut example of the global extinction of one species. But in the ranking or hierarchy of biological classification there are levels both above and below the species. Going to a lower level, one single species can be subdivided into various subspecies or varieties. These look similar overall, and they can also breed together. But they tend to live in separate geographical areas, split away from others of their species by mountain ranges, rivers or seas. In the wild they do not get the chance to interbreed, and so they follow their own evolutionary pathways as they adapt to their own particular environments. Some can be regarded as species in the making – or as subspecies disappearing.

Many subspecies have succumbed to extinction through the ages. The smallest tiger was the Bali tiger, *Panthera tigris balica* (a subspecies gets a third part to its scientific name). Found only on the Southeast Asian island of that name, it was small, for a tiger – at less than 100 kilogrammes (220 pounds), just half the bulk of the Bengal tiger on the Asian mainland. With nowhere to escape on its relatively small homeland, the Bali tiger was hunted to oblivion by the 1930s. In fact, three tiger subspecies have become extinct in the past century, leaving just six clinging on.

Lonesome no more

Much more recently, in 2012, Lonesome George passed away. Probably more than 100 years old, he died of natural causes. George was a Galapagos giant tortoise of the subspecies known as the Pinta Island tortoise, *Chelonoidis nigra abingdonii*. In 1971 George was rescued from devastating habitat destruction on Pinta, one of the smallest of the Galapagos group, in the East Pacific. Thereafter he was kept at the main Charles Darwin Research Station on nearby Santa Cruz Island. Huge efforts were made to locate a female of the same subspecies for George, anywhere in the world, so the two could mate and continue the line. But none was ever found. Meanwhile equally strenuous efforts were made to mate George with females of the most similar subspecies of Galapagos giant tortoise. Several of these females laid eggs. But none ever hatched. When Darwin himself visited the

● LONESOME GEORGE
George was named 'Lonesome' as the last of his kind – in his case the Pinta Island giant tortoise subspecies.

Weighing more than 200 kilogrammes (440 pounds), measuring 1.3 metres (4¼ feet) at the shoulder, and brandishing 1.5-metre (5-foot) horns, the giant sable antelope is an impressive but extremely rare inhabitant of scarce forests in Angola, southwest Africa. This formidable herbivore is a subspecies, *Hippotragus niger variani*, of the sable antelope species *Hippotragus niger*. Three other, smaller, subspecies live far to the east, separated by hundreds of kilometres.

The sable antelope as a species is officially International Union for Conservation of Nature (IUCN) 'Least Concern' – that is, no immediate danger of extinction. But the giant sable subspecies is categorized as 'Critically Endangered', at serious risk of extinction. Among its threats have been civil unrest and military conflict, including the long-running Angolan Civil War. It has been made the national animal symbol of Angola, with images on banknotes, postage stamps and the national airline; even the country's soccer team is called 'Palancas Negras' – the Black Sables.

● GIANT SABLE ANTELOPE
The male giant sable antelope has longer, thicker, more over-arching horns than the female.

Galapagos in 1835 there were probably 15 subspecies of giant tortoise. With the passing of George, 10 survive.

Rhinos as a group are the most threatened of all large mammals. The subspecies called the northern white or square-lipped rhinoceros, *Ceratotherium simum cottoni*, could hardly be rarer. Just a handful are left, a few in zoos and four relocated from captivity to a protected conservancy park in Kenya, Africa. Its prospects are bleak. In fact, recent studies show that this hulking herbivore may not be a subspecies at all, but a full species in its own right. The western black or hook-lipped rhino, subspecies *Diceros bicornis longipes*, was declared officially extinct in 2011.

Numerous other subspecies across the world are at imminent risk of extinction, from butterflies, fish and frogs to giant tortoises, birds, shrews, rabbits and giraffes. This dire situation can occur even when the animal is a treasured national icon of its region, such as the giant sable antelope of Angola (above).

Bigger than species

Above the species rank in biological classification are ever more encompassing groups. Related species (descended from a common ancestor and inheriting unique features) make up a genus. In turn, closely related genera form a family, then families combine into orders, and orders into classes. The most closely related classes come together into the largest, most encompassing groups, known as phyla for animals, or divisions for plants. (The modern system of grouping life-forms based on cladistic analysis is discussed later, see page 29.) Any of these rankings may go extinct, either as their last remaining species gradually peter out, or during a major event when the whole group, apparently established and thriving, rapidly disappears.

Among the most famous, long-lived and widespread of all prehistoric animal groups were the trilobites. They are usually given the rank of class, equivalent to their related groups such as insects and crustaceans (making up the phylum Arthropoda). Trilobites vaguely resembled a combination of crabs and woodlice and appeared in the seas some 520 million years ago. By 400 million years ago they were on the wane. A few types survived until the greatest-ever major extinction event at the end of the Permian period, 252 million years ago. Then this massive class of creatures – which during its time numbered some 10 orders, comprising over 150 families, containing around 5,000 genera, and totalling more than 15,000 and maybe even 20,000-plus species – went extinct.

Extinct from where?

As well as species or other groups coming to an end, there is also the consideration of where they disappear from. In a global extinction, the species is lost from the whole Earth. But there are also local and regional extinctions, known as extirpations. In these cases a species or other group no longer survives in a particular area or geographical region, but it is still present elsewhere. Local extinctions often happen to organisms that have a specific habitat or other needs. They exist in ecologically similar but widely separated zones or 'pockets' across their general range. These 'pockets' might be mountains separated by lowlands, islands divided by seas, or lakes with land between them. Since subspecies tend to evolve in these conditions, as described above, a local extinction often involves such a grouping.

● TRILOBITE
Once hugely varied and common, this vast group exists today only as fossils; this is a specimen of the genus *Dikelokephalina*.

Two well-known butterfly species involved in local extinctions from the British Isles were the large blue, *Maculinea arion*, and the large copper, *Lycaena dispar*. The latter went absent from British records in the 1840–50s, while the former held on until the 1970s. Both continued to live in continental Europe. A hope for local extinction is that the species can be reintroduced from surviving populations elsewhere. For the large copper this has been tried several times since the 1900s, bringing the subspecies *Lycaena dispar rutilus* from Germany and other European sites into Britain, and also *Lycaena dispar batavus* from the Netherlands, to replace the British native subspecies *Lycaena dispar dispar*. But without success. The original cause of the British extinction centred around drainage of fenlands and similar damp habitats for farming. Exactly why reintroductions have failed so far is not clear. But, as a general guide, if whatever caused a local extinction in the first place – such as habitat loss or pollution – is not fixed in the meantime, then reintroductions are unlikely to succeed.

The native British large blue, *Maculinea arion eutyphron*, was last seen in the UK in 1979. Conservationists keen to see it reintroduced carried out in-depth studies of its habitat needs and also clarified its amazingly complex life-cycle. The tiny newly hatched caterpillars feed at first on wild thyme. The final-stage caterpillar has a honey gland to make a sweet fluid, similar to plant nectar or aphid 'honeydew', that attracts ants. Certain kinds of red ants take the caterpillar to their nest, where it is accepted mainly because it produces a scent that mimics the ants' own larva. The caterpillar repays its hosts by eating their larvae (grubs) and growing enormously, before becoming a chrysalis (pupa) and finally the adult butterfly.

Knowing about these specialized needs allowed a series of reintroductions from Swedish populations of large blue, *Maculinea arion arion*, beginning in 1985. These have become established at several sites in western Britain, notably Somerset.

● LARGE COPPER
Several reintroduction attempts for the large copper butterfly into Britain, the most recent surviving until the 1990s, have all failed.

● LARGE BLUE
This spectacular butterfly is gradually regaining a foothold in Britain's West Country.

SPOTLIGHT ON... AMMONOIDS

The curly-shelled ammonites – more properly termed ammonoids – are some of the most familiar of all fossils. These sea-dwellers were cousins of today's octopus, squid and cuttlefish, within the great mollusc group. They first appeared some 400 million years ago and rapidly spread and diversified, although keeping the basic spiral shell shape. The ammonoids lived through several major extinction events, each time being reduced to a fraction of their previous numbers and variety, but then bouncing back and evolving again into yet more different kinds. Because they lived in the sea and had hard shells, they were excellent candidates for fossilization. And because they evolved rapidly into so many kinds, they are enormously useful. Their remains can be used as index (guide or marker) fossils to determine the age of the rock layers in which they occur, and so give a date to other fossils in that layer.

Ammonoids went into their final decline during the Cretaceous period, from 145 million years ago, and disappeared completely by the fifth major extinction event at the end of the Cretaceous, 66 million years ago. This was the same event that killed off so many other famous groups, including dinosaurs – well, most of them, as explained later... (see page 30).

● AMMONOID
Jeletzkytes nebrascensis from South Dakota, USA, was one of the last ammonoids, just before their extinction.

Extinction, time and fossils

Extinctions have several relationships with passing time. How long does a typical species last, from when it first evolves as a distinct biological entity to its final demise? How many extinctions might there be every, say, million years, either across the whole living world, or in various groups, such as flowering plants or mammals? Assuming a species is well established for a long time but then meets some great challenge, will its disappearance be relatively sudden (in geological terms) or long, drawn-out and tapering? Most of the knowledge to answer such questions comes from fossils. These are the remains of once-living species whose bodies, or parts of bodies, came to rest in sediments such as mud, silt or sand, and were very gradually replaced by rock-forming minerals and turned to stone. The fossil record over the vast span of geological time becomes progressively better-known with every year's batch of new discoveries. But the overall likelihood of any organism leaving any fossils is very low and the fossil record is riddled with chance events.

The result is that the fossil record is, to say the least, 'patchy'. For instance, it favours shelled or tough-cased animals in water. Shells are hard and more likely to persist long enough for preservation. Also, the aquatic habitat is more likely to bring sediments that cover the remains. Conversely, soft-bodied creatures such as worms in a tropical forest are almost certain to be devoured or rot away at speed, so these types of fossils are incredibly rare.

Scientists have ways of helping to compensate for these variations. For example, looking at the proportions today of shelled species in various watery habitats, compared to soft-bodied ones, provides a way of estimating what these proportions might have been in the past. As explained later, information about global air and sea temperatures, atmospheric oxygen and carbon dioxide levels, and many similar factors. is also taken into account. But it is still a speculative process, where almost everything is prefixed by terms such as 'estimated' and 'average'.

Also, the traditional definition of a species, which (in sexually reproducing species) involves individuals breeding together, cannot be applied to fossils. So distinct species in the fossil record are identified by sets or suites of physical features, such as numbers of spines or ridges on a shell, or the detailed shapes of teeth and bones in a mammal.

● TYRANNOSAURUS
This most famous prehistoric species probably existed for a few million years. Was this average for a dinosaur?

Background extinctions

Looking back over the fossil record, a very approximate average for the 'lifespan' of a typical species is 5–10 million years. Put another way, if there were around 5–10 million species in total on Earth – within an order of magnitude of most current estimates – then one species would go extinct each year. Or, put yet another way, the average is one extinction per million species-years. These types of educated guesses make up what is known as the background extinction rate. Of course, there are many variations with time and place. Fossils of the famed dinosaur species *Tyrannosaurus rex* crop up in rocks during a time span of around 3, possibly 4 million years. An average species of animal in the sea might last 4 or 5 million years. Then again, based on the fossil record of the past 65 million years, a typical species of the shellfish known as bivalves – members of the mollusc group, such as clams, oysters and mussels – could have lasted much longer, perhaps 8–12 million years.

The largest single group of animals on the planet is the insects, with around a million species described so far. This may well have been so in the past, too. But including insects in extinction rate estimates is notoriously difficult. Their small size, and liking for warm forests where natural recycling of bodies is rapid, mean their chances of leaving fossils are miniscule.

Five great extinctions

Background extinction rates refer to long periods of time, tens or hundreds of millions of years, when turnover of species 'jogs along'. New ones appear as old ones die away, at rates that fluctuate little. These low-level backgrounds are punctuated here and there by phases of mildly increased extinction rate, usually followed by a small burst of rapid evolution.

There were several occasions in the history of the Earth where something even more dramatic happened: geologically short time spans (thousands to a few million years) during which a large proportion of species became extinct. These were major extinction events or 'mass extinctions'. They are named according to the geological periods when they occurred. Geological periods are the main units of time making up the immense history of Earth. Each period spans tens of millions of years and is defined by the types of rocks formed while it lasted. In many cases the end of a period is signalled by a marked change in the types of rocks laid down, because of vast environmental transformations on Earth – which is why there were also great changes in life.

Most scientists recognize five such mass extinction events. The more distant they are in time, the less detailed evidence we have about them today – their fossils were destroyed as rocks were worn away, cracked open, heated or even melted, then reformed, and so on, many times through the ages. Unravelling the causes of these five events (discussed later, see pages 16-25) has great relevance today, not only to understand prehistoric extinction events and evolutionary processes, but also to help shape our views on the burning issue of a current sixth mass extinction.

- Ordovician–Silurian, 445–440 million years ago

   Two spates of extinctions, with 1 to 2 million years between, caused what was probably the second-greatest loss of biodiversity on Earth. These happened as the Ordovician period gave way to the Silurian, at a time when life on land was just becoming established and fishes were coming to dominate the seas. Perhaps one-quarter of all biological families perished, and more than half of all genera (groups of species), in this end-of-Ordovician or O–S event.

● DUNKLEOSTEUS
Even giant top predators such as *Dunkleosteus* could not beat the rigours of the Late Devonian major extinction.

● Late Devonian, 375–359 million years ago

As the Devonian period came to a close, there was another major extinction event. The Devonian was the 'Age of Fishes', so the fossil evidence is from sea animals, although plants, and insects and similar creatures, were spreading over the land. One major fish group that met its end was the placoderms, including the giant, razor-jawed, seemingly indestructible predator *Dunkleosteus*, 10 metres (33 feet) long and weighing 3 or 4 tonnes. The timescale of the end-of-Devonian or D–C (Carboniferous) event is not very clear, with possibilities ranging from a few sharp peaks of losses in less than 5 million years, to a longer, drawn-out series of accumulative losses over 15–20 million years. An estimated one-half of all genera disappeared, and some assessments suggest three-quarters of all species were lost.

● Permian–Triassic, 252 million years ago

The end-Permian or Permo-Triassic (P–Tr) extinction event was the big one – the greatest loss of living things and biodiversity the world has ever seen. It is referred to colloquially as the 'Great Dying'. More than half of all biological families met their end. With life on land well established, there are estimates for marine, freshwater and land-based organisms. In terms of animal species, perhaps two-thirds of those on land (insects included), and amazingly more than nine-tenths in the sea, met their end. Recent studies indicate that the P–Tr extinction's main phase could have lasted just less than one-quarter of a million years (see pages 18-19).

● Triassic–Jurassic, 201 million years ago

The end-Triassic or Triassic-Jurassic (Tr–J) mass extinction involved around one-quarter of animal families, encompassing one-half of genera and probably a similar proportion of species. Some evidence suggests it was very rapid, lasting less than half a million years, and perhaps as little as 10,000 years. Groups such as the large amphibians, previously dominant in various habitats, came to an end. This opened the door for another group that soon came to rule the land – the dinosaurs.

● Cretaceous–Palaeogene, 66 million years ago

The end-Cretaceous major extinction was long known by shorthand reference as the K–T event – K for *kreide* or *creta* (chalk, Cretaceous), and T for the Tertiary period that followed the Cretaceous. However, some years ago scientific convention split the Tertiary into two periods, the Palaeogene (66 to 23 million years ago) and the Neogene (23 to 2.6 million years ago). So the more modern (though less snappy-

SPOTLIGHT ON... HORSESHOE CRABS

The sea creatures known as horseshoe crabs or king crabs are among the greatest of all animal survivors. They are not true crabs (which are crustaceans) but belong to a group named the xiphosurids, their cousins being spiders, scorpions and other arachnids. The four species alive today, in the genera *Carcinoscorpius*, *Limulus* and *Tachypleus*, are popularly known as 'living fossils'. They closely resemble (but are not the same species as) their relatives from Jurassic times, more than 150 million years ago. As a group, horseshoe crab history goes back much further, some 445 million years. This means the seemingly unassuming xiphosurids have cheated four, if not all five, major extinction events.

Today's horseshoe crabs grow to a maximum length of 60 centimetres (2 feet) including the long tail spike. They have a strong all-over shell, the carapace, for protection, also the ability to regrow lost limbs, and an adaptable diet of worms, shellfish, other seabed creatures and general debris. Perhaps these are all factors in their staying power. More long-term survivors are described on later pages (see, for example, pages 34-37).

● MESOLIMULUS HORSESHOE CRAB
Fossilized *Mesolimulus*, 150 million years old, is hardly distinguishable from its living cousins.

sounding) shorthand name is the K–Pg event. This most famous mass eradication wiped out half of genera, and perhaps three-quarters of all species. They included big, spectacular beasts such as mosasaurs and plesiosaurs in the sea, and pterosaurs in the air. On land, the dinosaurs were also involved. The causes of this catastrophe are discussed later (see pages 26–29).

Nature bounces back

Each major extinction event was followed by a burst of evolution as new species adapted to fill the many previously occupied but now 'vacant' ecological roles, or niches. This rebound is discussed later (see page 26). So for extinctions through the ages, whether at a steady background rate or during a major event, a mix of destructive and creative biological processes meant that the end for some provided great new opportunities for others.

driven to disappear
CAUSES OF EXTINCTION

WHAT DRIVES A SPECIES or other group to disappear from the Earth? There are many, many causes – global, regional, local, clear-cut or complex, coming from outer space or generated here on Earth. Sometimes there is a single main cause, but more often it is a combination of factors. The slow crawl of continents around the Earth's surface, known as tectonic plate drift, is global in scale. As this happens the oceans change shape and their water currents alter, and landmasses come together or spread apart. Sometimes these changes reach a 'tipping point' and lead to accelerating climate change. For example, as two continents meet, they close a channel that was important for ocean currents to spread warm waters around the world. Now that can no longer happen. Possibly too, the colliding continents buckle and begin to throw up a mountain range, which likewise alters atmospheric circulation. Over a relatively short period of time, climate patterns are transformed. Wind and rainfall alter, their bands and zones move, global temperatures go up or down, sea levels rise or fall.

● HENBURY METEORITE
The 28-cm Henbury iron meteorite was once in an asteroid's core; asteroid impacts may have contributed to mass extinctions.

Easier to comprehend, and also huge in scale, are sudden catastrophic great events such as asteroid and meteorite impacts, and massive, widespread volcanic eruptions. These have speedier effects, such as pouring dust into the atmosphere that blots out the sun and brings a period of rapid cooling.

Earth in space

Even the motion of planet Earth through space has effects on life. Our globe precesses or 'wobbles' on its axis, like a fast-spinning toy top that leans at an angle which turns a slow circle. The main cycle of Earth's precession occurs over a period of 26,000 years. This wobble alters the amount of the sun's heat and light received by different parts of the surface. In addition, the Earth varies slightly in its distance from the sun, which is known as orbital eccentricity. When the orbit becomes more elongated or elliptical, rather than more circular, the seasonal variations are more extreme.

All of these factors affect living conditions and the ability of plants and animals to avoid extinction. The following three scenarios – one from very ancient times, the second from just a few million years ago, the third from only thousands of years ago – may help to illustrate just a few of the complexities involved.

● EYJAFJALLAJÖKULL ERUPTION
Volcanic activity – here on Iceland in 2010 (which shut down European air travel for days) – is implicated in several major extinction events.

SPOTLIGHT ON... GIANT DRAGONFLIES

The largest winged insects known arose during the Carboniferous period, more than 300 million years ago. They are usually called giant dragonflies and included genera such as *Meganeura* and *Meganeuropsis*. However, they are not true dragonflies, rather they are ancient cousins, and sometimes called griffinflies. The biggest had wingspans of more than 75 centimetres (30 inches). Possibly the high levels of atmospheric oxygen during the Carboniferous, coupled with a slightly

● GRIFFINFLY
'Giant dragonflies' like *Meganeuropsis* were the biggest flying creatures 300 million years ago.

different breathing process from other insects, helped these fliers to evolve to such great sizes.

Griffinflies flourished during the Carboniferous period and some continued into the following Permian. But they faded and met their end towards the close of the Permian.

What led to the 'Great Dying'?

Several causes are suggested for the greatest-ever major extinction, the end-Permian event, 252 million years ago. Researchers look for clues in the fossils of the time, the rocks containing them, and the patterns of extinctions, which affected some groups of organisms severely but others less so.

A leading contender involves one of the greatest ever volcanic events on Earth. This was molten rock, lava, flooding from the Siberian Traps in northern Asia, to cover an area of more than two million square kilometres. Geological evidence reveals that the eruptions also spewed out vast amounts of ash and other debris, which would spread in the atmosphere, reduce sunlight around much of the world, and hinder plant growth. Acidic fumes from those volcanic emissions may have generated thousands of years of acid rain, while the potent greenhouse gas methane gushed from the same source, leading to tremendous challenges for plant growth and life in general. The date of the Siberian Traps eruptions coincides remarkably well with the end-Permian event.

The Great American Interchange

Many factors can bring about extinctions. They include geographical isolation, droughts, floods, heatwaves, cold snaps and mini ice ages – all events on a vast scale in the non-living environment. There is another whole range of factors concerning the living surroundings, and interactions with other organisms. They include being out-competed by newly evolved or invading species for resources such as food, homes or living space; falling victim to new or improved predators; losing a major food source because that has become extinct; and succumbing to epidemics of disease spread by new strains of microbes. These life-based or biotic factors were important in a series of intercontinental invasions and extinctions within the past few million years, in the Americas.

About 3 million years ago, after a period of isolation lasting some 30 million years, South America became joined to North America. Tectonic drift and sea-level changes saw the formation of a land bridge, the Isthmus of Panama. Terrestrial creatures and plants were able to move from the south to the north, or vice versa. Known as the Great American Interchange, this event had far-reaching effects on biodiversity for both continents. At the same time, communication between the Pacific and Atlantic Oceans ceased, altering sea current patterns and climate in the region.

● ASTRAPOTHERIUM
Now-extinct astrapotheres showed similar features to modern tapirs.

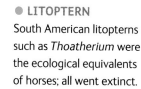

● LITOPTERN
South American litopterns such as *Thoatherium* were the ecological equivalents of horses; all went extinct.

Much as today on the island continent of Australia, South American life had been evolving in 'splendid isolation'. There were many unique kinds, including those that filled the ecological roles or niches occupied by different groups in other parts of the world. Among the hoofed mammals, litopterns were browsers and grazers, leading lifestyles similar to the horses, camels, antelopes and deer elsewhere. Pyrotheres and astrapotheres resembled elephants and tapirs in other regions (see previous page). Nototheres or 'southern ungulates' were a very varied group, some similar to rabbits while others were heavy-bodied, more like hippos. The process whereby separate groups of living things evolve to have similar features and lifestyles, because they occupy similar habitats and ecological roles, is known as convergent evolution.

As the Great American Interchange picked up speed, many South American species went extinct. They were out-competed and ousted by invaders from the north. For the hoofed mammals, these included horses, deer, and members of the camel and tapir groups. Far fewer migrants went from south to north and made the big time. There are various explanations for the northerners' success. Compared to South America, the continent of North America had a wider range of habitats, and also more episodes of being joined to other landmasses, with much more interchange and fiercer competition. The result of the interchange was that thousands of South American species disappeared.

Battle of the sabre-tooths

One fascinating aspect of the Great American Interchange concerns two 'sabre-tooths'. *Smilodon* was a true cat, or felid, that appeared in North America more than 2 million years ago. As it spread across the land bridge into South America, larger *Smilodon* species evolved.

But the 'sabre-tooth' feature has evolved separately in various predators through the ages, and in South America there was already one – *Thylacosmilus*. Outwardly, this

● **THYLACOSMILUS**
Despite its formidable curved fangs, *Thylacosmilus* lost out, probably in part to rival *Smilodon* from the north.

SPOTLIGHT ON... SMILODON

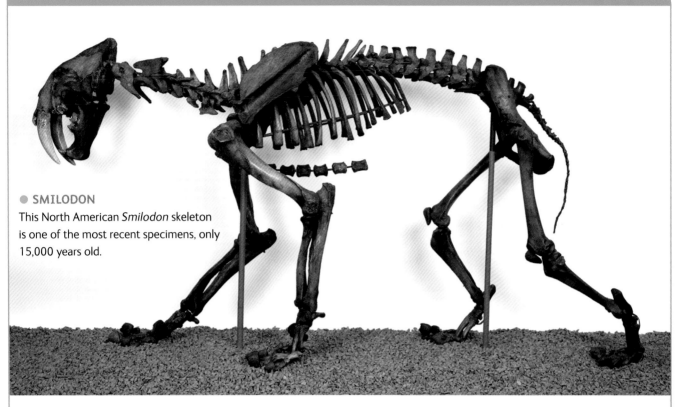

● SMILODON
This North American *Smilodon* skeleton is one of the most recent specimens, only 15,000 years old.

In North America 2 million years ago, the sabre-tooth cat species *Smilodon gracilis* weighed up to 100 kilogrammes (220 pounds), roughly the same as today's American big cat, the jaguar. This fearsome prehistoric predator went extinct perhaps half a million years ago, but not before it or a close relation evolved into a larger version, *Smilodon fatalis*. This species arrived on the scene 1.8–1.6 million years ago and was about the same size as the largest present-day cat, the Siberian tiger. Fossils show how it spread as part of the Great American Interchange into South America. Once there, again it or a close relation evolved, by 1 million years ago, into an even larger species, *Smilodon populator*.

With a shoulder height of 1.5 metres (5 feet), a weight of more than 400 kilogrammes (970 pounds), and curving sabre-like canines 30 centimetres (12 inches) long, *Smilodon populator* was one of the most massive cats ever. However, by 10,000 years ago the last *Smilodon*, both in South America and back in the north, were gone. Perhaps humans, along with climate change, induced extinctions in *Smilodon*'s prey species, which in turn resulted in the extinction of the top predator.

150-kilogramme (330-pound) cat-like hunter resembled *Smilodon*. But it was from a very different group of mammals known as sparassodonts, more closely related to marsupials, such as today's kangaroos, koalas and opossums.

Thylacosmilus was a formidable predator. But, as the Great American Interchange took a grip, it could no longer hold its own. As mentioned above, northerners like *Smilodon* were generally equipped with more recent, often more effective evolutionary

● GIANT ELK
The giant elk's vast antlers were shed and regrown yearly, requiring a huge nutrient supply.

The giant or 'Irish' elk, *Megaloceros giganteus*, was certainly a giant of its kind. But it was not exclusively Irish. Neither was it a true elk, in the sense that it was not in the same deer group as the modern-day North American moose or European elk, *Alces alces*.

This massive herbivore made its appearance some 400,000 years ago. Gradually it spread across much of northern Europe and northern Asia, with remains found from Ireland in the west, to the wilds of Siberia in the east. About 2 metres (6½ feet) tall at the shoulder, it weighed in excess of 600 kilogrammes (1,300 pounds). The male's extravagant antlers made up 40 kilogrammes (nearly 90 pounds) of this, and from tip to tip their span approached 4 metres (13 feet). The vast antlers may be a result of the well-known biological phenomenon known as allometry. This involves certain body parts increasing in size at a greater rate, in proportion to the increase in size of the overall body. So as the elk's main body got bigger, perhaps as an adaptation to ice age cold, the antlers increased in size even faster.

Giant elk remains are known from the preserving peat bogs of Ireland (hence 'Irish' elk), with the most recent remains dating to 10,000 years ago. Specimens from western Siberia are slightly younger, less than 8,000 years old. As well as over-hunting by humans, other factors are suggested for the giant elk's demise. Climate change may have altered the vegetation, so that the evolving plants could no longer provide the minerals needed to grow such a large body and outsized antlers.

features. Whether the two great hunters ever went head-to-head, or sabre-to-sabre, is not known. But perhaps *Smilodon* was more efficient at finding, tracking and bringing down victims. Once *Smilodon* became established, it undoubedly took many native South American prey animals that were also part of *Thylacosmilus'* diet, and so contributed to both their and *Thylacosmilus'* extinction.

Ice age extinctions

Earth has become much warmer or far cooler hundreds of times through its immense history. These changes have been prime drivers of evolution and extinction. The last series of glacial periods, or Pleistocene ice ages, began more than 2 million years ago. Within these, average global temperature drastically rose and fell several times. The recent dip – the most recent glacial period – started some 100,000 years ago. It reached its low, with ice-sheets and glaciers at the greatest extent as the LGM, Last Glacial Maximum, from around 25,000 to 13,000–10,000 years ago. This is the popular concept of the 'last Ice Age'. Since then the planet has warmed. But Earth's history is so immense that, even with our current situation, there is no guarantee the LGM will be the 'last', or that the Pleistocene ice ages are finished.

These changes were fast by the scale of geological time, with the ice front advancing as rapidly as 500 metres (1,650 feet) per year, and then retreating during the interglacial. Causes of ice ages probably included a complicated mix of: tectonic drift, ocean currents, atmospheric oscillations, even Earth's periodic movements in space. Our planet's angle to the sun, daily spin and yearly orbit are not regular. There is a 26,000-year cycle or 'wobble' of the Earth about its spin axis. In addition, the overall angle of the spin axis, known as obliquity, changes by a few degrees over a 41,000-year cycle. These and other variations combine in different ways at different times to affect temperatures and climates. For the last batch of ice ages, we know a great deal about climate and wildlife changes because fossils and other remnants, such as frozen carcasses in glaciers, and mummified remains and dung in caves, are plentiful. And they reveal plenty of extinctions.

Megafauna no more

The loss of megafauna over the past 40,000–10,000 years is striking. Fauna are the animals that live in a particular time and place; megafauna are the really big ones, mainly mammals. As though great coordinating forces were at work, during that period many large mammals across much of the world died out. This was a period of climate change across the globe as the ice-sheets spread down from the Arctic across many northern lands, then retreated. Whole habitat zones with their plants and animals shifted south in front of the ice, then back north with its retreat.

Across the world, hundreds of populations of large mammal species disappeared. Some correlate with shifting climate patterns. Maybe they could not migrate fast enough, or evolve rapidly enough, to keep up with the environmental changes. Perhaps they were out-competed by rivals fleeing from harsher conditions elsewhere. But many

● GIANT GROUND SLOTH

The giant ground sloth *Megatherium* was probably extinct by 10,000 years ago. Its fossils were collected by Charles Darwin on his HMS *Beagle* voyage, 1831–36.

of these extinctions have little or no link to climate change. Some of them do, however, correlate with the spread of humans around the globe.

Southern extinctions

No continent was exempt from these megafaunal extinctions. Some 50,000–40,000 years ago they were happening far from the ice-covered northern lands, in Australia. Just a few of the dozens of eventual victims were the 'giant wombat', *Diprotodon*, the 'marsupial lion', *Thylacoleo*, and several kinds of huge kangaroos, such as *Procoptodon*. It is usually accepted that humans had arrived in Australia by 40,000 years ago.

In South America, some megamammals had survived the Great American Interchange. But they were hit by a spate of extinctions, especially around 12,000–10,000 years ago. The giant ground sloth, *Megatherium*, standing 5 metres (16½ feet) tall and weighing more than 5 tonnes, was one famous casualty. Another was the litoptern (hoofed mammal) *Macrauchenia*, a vaguely camel-shaped beast with a long trunk-like snout. It probably weighed around a tonne. By this time, humans with efficient killing weapons such as spears had spread into this continent from the north.

Across the north

North America has a long list of big mammals that also underwent a rash of extinctions in South America, beginning around 12,000 years ago. They include not just herbivores but also predators and omnivores: Columbian and woolly mammoths, *Mammuthus columbi* and *M. primigenius*; American mastodon, *Mammut americanum*; ancient bison, *Bison antiquus*; the American horse, *Equus scotti*; the 100-kilogramme (220-pound) giant beaver, *Castoroides ohioensis*; a short-faced bear, *Arctodus simus*; the dire wolf, *Canis dirus*; sabre-tooth cats including *Smilodon*, and dozens more. About this time, the northern ice-sheets retreated and habitats across the continent changed. Also, from about 13,000 years ago Clovis people lived across the continent and fashioned Clovis points – sharp-edged stone blades used to tip spears, arrows, darts and knife shafts, for efficient killing and butchering of large carcasses.

Megafauna in Europe and most of Asia suffered similar fates: woolly mammoth; woolly rhino, *Coelodonta antiquitatis*; giant elk; steppe bison, *Bison priscus*; cave lion, *Panthera leo spelaea*. Around this time humans in Europe were moving from using simpler stone tools and weapons to the more sophisticated versions of the Mesolithic or Middle Stone Age. There were short-term climate fluctuations during these torrid times, which may have put more pressure on already stressed large mammals. They include the Older and Younger Dryas, two brief, very cool and dry periods from 18,000–15,000 and 12,800–11,500 years ago respectively. The Younger Dryas, in particular, interrupted the gradual warming trend in Europe. These three extinction examples: end-Permian, Great American Interchange, and Ice Age megafauna loss, took place during very different phases in Earth's history. Yet probably common to all three were climate change. Humans were absent in the first two but probably helped the third.

● PROCOPTODON

This giant short-faced kangaroo stood 2 metres (6 ½ ft) tall. Dating its most recent remains is controversial, ranging from 46,000 to 15,000 years ago.

● CLOVIS POINT

The Clovis point dates from around 13,500 years ago.

creative forces
EXTINCTION AND BIODIVERSITY

ONE FEATURE WAS COMMON TO ALL major mass extinction events: nature eventually recovered. After an interval, new species arose rapidly by many means to occupy 'empty' sets of ecological needs, or niches. New grazers and browsers arose to replace extinct ones, new predators appeared to hunt them, and so on. This is the positive side of the extinction coin, as a drastic loss in biodiversity is followed by a rebound increase. It usually took several million years for life's variety to recover to something like pre-event levels, and after the most serious, the end-Permian event, it took 10–20 million years.

The best-known example of biodiversity's rebound from a mass extinction, partly because it was the most recent, and so we have the most evidence, was the end-Cretaceous extinction 66 million years ago. Dinosaurs had ruled the land for more than 150 million years, during the preceding Jurassic, and the Cretaceous itself. In most land habitats they were by far the most common large animals.

No single dinosaur group survived all through this immense time span. But dinosaurs as a whole stayed in charge as different types arose from within the ranks and then went extinct. For example, the sauropods – long-necked and long-tailed plant-eaters, like *Diplodocus* and *Brachiosaurus* – had their heyday at the end of the Jurassic and Early Cretaceous. Then in many regions they faded as other groups, such as the hadrosaurs or 'duckbills' in North America, took over the role of chief plant-eaters. These types of replacements followed the usual evolutionary patterns. They were driven by continuing changes in the surroundings, like a drier, cooler climate, and, during the Cretaceous, by the spread of a new plant group as food, such as the angiosperms or flowering plants.

The impact theory

Some surveys show that dinosaurs as a whole were already on the wane by the end of the Cretaceous. Then came the sudden extinction event. This is usually attributed to a massive space rock, perhaps an asteroid 10 kilometres (6¼ miles) across, which hit the Earth in the region of Yucatan, Mexico, and created the Chicxulub Crater. Also, rocks formed at this time around much of the world show unusually high levels of the chemical element iridium. This element is rare on Earth – but common in space bodies. Rocks laid down at the Cretaceous–Palaeogene transition had 10–15 times more iridium than expected.

SPOTLIGHT ON... CHICXULUB CRATER

This giant 'dent' in Mexico's Yucatan Peninsula and the neighbouring ocean was more than 180 kilometres (110 miles) across and up to 10 kilometres (6¼ miles) deep. It is named after the town of Chicxulub, near its centre, and was first noticed by petroleum surveyors in the 1970s. The date of the crater's origin fits well with impact theories at the close of the Cretaceous period.

The original shape of the crater is now obscured by subsequent earth movements, erosion, and filling by limestone sediments and general seabed mud. Recent radar images from space reveal its remnants, including a 'trough' some 3 to 5 kilometres (2 to 3 miles) wide that arcs around the crater's contour, marking structural disturbance in the rocks beneath. Chicxulub is now regarded as good evidence for the asteroid impact. It is more controversial whether this impact essentially caused, or partly contributed to, the end-Cretaceous mass extinction.

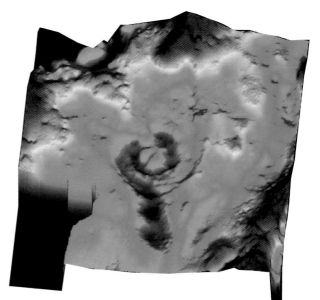

● CHICXULUB CRATER
A computer-coloured gravity map shows the Chicxulub crater (blue, centre) filled with low-density impact debris and sediments.

The theory runs that a massive impact caused vaporization of the space rock as well as the Earth rocks it hit. Material spread into the atmosphere, gradually settled and became incorporated into new rock formations. Recent evidence from craters in the North Sea, Ukraine and Indian Ocean suggests there could have been multiple impacts.

Asteroid winter

What happened before and after the end-Cretaceous impact(s) remains speculative. Prior to this time, enormous volcanic activity was going on in the region known as the Deccan Traps, in India. Ash, vapours and toxic gases would be pouring into the atmosphere. Then the space-rock impact threw up dust and other materials too, further blocking the sun's light and heat. This is the popular concept of an 'asteroid winter'. Plants withered in the cool gloom, herbivores went hungry and starved, and so did the predators who hunted them. Other possible processes include widespread earthquakes, and acid rain from rock minerals dispersed into the atmosphere. If vast wildfires took hold and sent smoke, ash and especially carbon dioxide into the air, this could even have triggered a short-lived 'greenhouse effect'.

Some scientists support the view that the extinction was more gradual. Climate change was already well under way towards the end of the Cretaceous, with a significant fall in sea levels, known as marine regression. Ocean currents were altering. Maybe the impact was something of a 'final straw'.

The whole scenario for the end-Cretaceous event continues to be widely debated. To many minds the pattern of extinction is puzzling, with large animals on land perishing, but not those in watery environments. Whatever the details, the event had many consequences. Numerous groups of animals and plants were wiped out. The 'empty' habitats were ready to be repopulated with new species, the ancestors of which lived side-by-side with the dinosaurs.

● DECCAN TRAPS
Much of this half-million square kilometres (almost 200,000 sq miles) of volcanic rock, known as flood basalt, formed in less than 50,000 years.

Our scientific view of extinction

Ever since the dinosaur group was officially named in 1842 by eminent British anatomist and fossil expert Richard Owen, it was understood that all of these great reptiles were extinct. But scientific views have changed. The way living things are grouped and classified has altered in recent years. Scientists now use an approach termed cladistics. This is based on units or groupings called clades. A clade consists of an ancestor organism and all its descendants, but nothing else. This applies to the living and the extinct. The clade members have features or traits which are unique, that is, no other clade has them, and which they share because they were inherited from the common ancestor. The system allows a clade to contain one or more further clades, provided members of each further clade have their own unique, derived features or characters. In this way clades 'nest' together and form a hierarchy of classification.

Dinosaurs are regarded as a clade. They share a suite of features that they all inherited from their common ancestor, and which no other reptiles or other creatures have. These include particular gaps in the skull bone, a crest on the shin bone, a certain arrangement of ankle bones, and other skeletal details seen in fossils.

Dinosaurs are not extinct!

It is generally agreed that some time during the Late Jurassic period, before 150 million years ago, birds evolved from a group of smallish, meat-eating dinosaurs. So birds lie within the dinosaur clade. Indeed, they should be regarded as dinosaurs. Birds are obviously far from extinct, sharing the modern world with us. So, if birds are living, and birds are dinosaurs, then dinosaurs are living – not extinct. This may seem counter-intuitive or even ridiculous to some people. But consider this analogy. Motor vehicles have 'evolved' from their simple beginnings into so many different kinds today – family saloons, workhorse vans, huge trucks, racing cars and so on. Imagine all motor vehicles except racing cars suddenly disappeared. We would then say saloons, vans and trucks were extinct. But we would not consider that all motor vehicles are extinct, because they are represented by survivors, in the shape of racing cars. So not all dinosaurs are extinct, because they are still represented by one of their surviving sub-groups, in the shape of birds.

Of course, creatures like *Tyrannosaurus* and *Triceratops* are truly extinct. It is the modern approach to classification, systematics and biology in general that has, in a sense, 'resurrected' them. Phrases about what happened to dinosaurs at the end of the Cretaceous now go something like 'non-bird dinosaurs disappeared', 'non-avian dinosaurs died out' or 'most dinosaurs met their end'.

● REPENOMAMUS
The largest Dinosaur Age mammal was *Repenomamus* from 130 million years ago. Its fossils suggest it ate small dinosaurs.

New opportunities, fresh challenges

After major extinctions such as the end-Cretaceous event, with a drastic loss of numbers and variety of life, there were fresh surges of evolution after an initial recovery interval. This process is known as adaptive or evolutionary radiation. During this time of almost frantic evolutionary activity, speciations – the origin of new species – proceeded at a rapid rate. Biodiversity was being restored. Surviving animal groups, which may have been relatively conservative in their evolution for a long time before the event, quickly began to adapt and take up new roles. It can be seen as a time for 'experiments in evolution'. Some soon failed, while others laid the foundations of success.

The disappearance of most dinosaurs meant there were vacant ecological niches – positions or roles to be played within the ecosystem, such as a large swamp-dwelling herbivore, a medium-sized desert predator, and so on. Overall, the places of the non-avian dinosaurs which went extinct were taken by mammals. Mammals had been around almost as long as the dinosaurs, but they could not compete in terms of size. The largest mammal throughout the dinosaurs' reign, known from adequate fossil evidence, was the badger-sized carnivore *Repenomamus robustus*, about 1 metre (3⅓ feet) long and 12 kilogrammes (26½ pounds) in weight. Most other mammals were similar in size and overall looks to today's mice, rats, shrews and weasels. Indeed the general 'shrew body plan' of a small, nocturnal, insect-eating mammal has coped well through the ages. Several different mammal groups evolved to fill this mini-predator niche, survived the end-Cretaceous event, and continue to the present day.

● SHREW
Most mammals of the Dinosaur Age were the size of, or not much larger than, modern shrews.

Big and bizarre

Within a few million years of the end of the Cretaceous, new mammals evolved that were both big and bizarre (to our modern eyes). One group was the mesonychids. Now extinct, they were related to hoofed mammals such as today's deer, antelopes, pigs, sheep and cattle. But with the big hunting dinosaurs gone, mesonychids took the opportunity to evolve into carnivores. For example, *Dissacus* appeared within a million years of the extinction. It resembled a wolf or jackal, occupied the role of medium-sized predator, and spread to most northern lands. A couple of million years later came the even bigger, fiercer *Ankalagon*. However, this too was extinct just a few million years later, as both its prey and its competing predators made their own evolutionary advances.

Birds, too, were in on the act. Some became outsized and adapted for the hunter-ambusher-scavenger lifestyle previously occupied by their dinosaur kin. By 55 million years ago tiny-winged, flightless *Gastornis* stalked Europe and North America. Up to 2 metres (nearly 7 feet) tall, muscular and with powerful legs, it had an enormously deep, strong beak that seemed ideal for ripping open prey or carcasses.

The evolutionary race continued as new 'improved' species regularly appeared to take over the niches occupied by older, less well adapted ones. About

● GASTORNIS
Post-Cretaceous landscapes saw various giant, flightless, carnivorous birds. Here *Gastornis* seizes *Hyracotherium*, a horse-like animal but only the size of a pet cat.

SPOTLIGHT ON... ANKALAGON

No sooner were big predatory dinosaurs gone than several mammal groups evolved quickly to fill their niches. Big as a modern bear, but with the general shape of a wolf, *Ankalagon* was one of the largest mammals in the aftermath of the mass extinction. By 63 million years ago it was already stalking North America. It had powerful jaws, curving canine teeth, and massive ridged molars with a shearing–crushing action, similar to today's hyaena. Some fossil jaws and teeth of *Ankalagon* are larger than others, suggesting that one sex was

● ANKALAGON
This formidable predator appeared soon after the end-Cretaceous event, but was gone itself 3 million years later.

larger than the other. Looking at these types of size differences in similar creatures, it is likely that the male *Ankalagon* was larger, with teeth better equipped for crushing carcasses and so filling the niche of large mammal scavengers. This is another example of sexual dimorphism, as mentioned earlier (see p.6).

20 million years after the end of the Cretaceous, during the Eocene epoch, many of the modern mammal groups which had originated during the Cretaceous were now becoming much more widespread and established: ungulates (hoofed animals), elephants, rodents, bats and our own group, the primates (lemurs, monkeys and apes). A great extinction had once again led to a new diversity of life on Earth.

Extinct no more

We have already mentioned dinosaurs 'coming back to life', as a result of the shift in scientific terminology (see pages 29-30), but occasionally life sciences see a very different resurrection. An animal or plant that was long believed to be extinct reappears. This is known as a 'Lazarus species', from The Bible's account of Lazarus miraculously being raised from the dead.

It might happen that a species from historical times has been missing, with no good evidence of its existence (such as confirmed sightings or bodily remains) for a considerable time. The time span varies according to the species and its predicament, but in general there should be no reasonable doubt that the last individual has died; this, in turn, relies on exhaustive surveys in its known and suspected habitats, at appropriate times. Suppose all these criteria are fulfilled — then the species is unexpectedly found again in a remote location such as an isolated mountain valley or a deep lake. This might be for a simple reason, such as not enough knowledgeable people looking in suitable places. Or the species might have disappeared from its usual range and migrated to, or hung on in, a place where no one expected to see it.

Insects, fish and frogs, lizards, birds like the forest owlet and Madagascar serpent eagle, and mammals ranging from Gilbert's potoroo or rat-kangaroo, and the very cute Santiago Galápagos mouse to the Philippine naked-backed fruit bat — all have come back from the presumed dead.

Gone, back, going again

Each rediscovery has a fascinating story to tell. In Australia's Victoria state Leadbeater's possum, *Gymnobelideus leadbeateri*, a tree-dwelling marsupial, was presumed extinct from 1909, its last confirmed presence. In 1961 surprised naturalists saw and then trapped the possum again, in less frequented upland forests of the region. The habitat has to be very precise for this possum's narrow ecological needs, or niche. The possum became a celebrated cause, and conservation strategies swung into action. Then in 2009 the horrific 'Black Saturday' bushfires burned a swathe of destruction across Victoria state, including most of the possum's highly restricted range. Many of the possums perished. Desperate measures were taken to capture and breed a few survivors. Having been 'extinct', then extant, the possum is once again on the way out.

In India, the small forest owlet, *Athene blewitti*, was first described for science in the 1870s. There are records of sightings in the next decade, then — nothing. More than a century later, in 1997, the species turned up during surveys some distance

● FOREST OWLET
Locations fraudulently recorded as sightings by a now-discredited bird collector misled experts, who searched in these wrong places, found no owlets and so considered the species extinct.

Believed actually extinct for so long, the African coelacanth was famously rediscovered in 1938. It is now IUCN-classed as 'Critically Endangered' – just one category away from 'Extinct' (again).

from its original location. Further investigations since have found it at several scattered sites across central India, although in all of these, it seems to be rare. The species is listed by the IUCN as 'Critically Endangered'. There is a strange twist to this owl's story. The first specimens for scientific description were submitted to London's Natural History Museum by British military officer and bird expert Richard Meinertzhagen. Later, detailed investigations of Meinertzhagen's army career and his nature-collecting suggest that he was far from an honest character. It appears that he stole bird specimens from other collectors, relabelled them with new false information, and submitted them as his own new discoveries. The faked localities he described for 'his' forest owlet probably distracted searches for more than a century, leading it to be erroneously categorized as extinct.

Back from extinction, again

Towards the other end of the timescale for Lazarus species, far away from reappearance in a century or two, a plant or animal may disappear from the fossil record for millions of years – then return. It may recur as a fossil again, or even turn up living and breathing.

One of the best-known examples is the fish group known as coelacanths. These were last known to science as fossils from the Late Cretaceous period, about 70–66 million years ago. These fish, in turn, were not greatly changed from the originals of the group far earlier, 400 million years ago.

In 1938 the scientific world was stunned by the capture of a living coelacanth, *Latimeria chalumnae*, along the southeast coast of South Africa. Local people knew this fish as an occasional and unusual catch, and there were reports that they used its rough skin as sandpaper, for example when roughening bicycle inner tubes before sticking on

a puncture-repair patch. But science did not know about the coelacanth until the 1938 find. The African or West Indian Ocean coelacanth rapidly gained world renown as a 'living fossil', although it was actually a different species from its prehistoric relations. Nicknamed 'Old Four Legs', it was especially important because its fleshy-based fins give clues as to how related fish, from the group known as sarcopterygians, may have gradually evolved their fins into legs and walked on land.

Further coelacanth specimens have come to light around eastern South Africa and the West Indian Ocean. Then in 1997 – another amazing chance find, and again in a local fish market. A coelacanth was discovered in Sulawesi, Southeast Asia. It is now known as the Indonesian coelacanth, *Latimeria menadoensis*. So the coelacanth was raised from the grave of extinction, not once, but twice.

The 'dinosaur pine'

Plants have also been re-found after millions of years of absence, presumed extinct. The Wollemi pine, *Wollemia nobilis*, astounded botanists when in 1994 it was described from a series of rainforested steep valleys, less than 200 kilometres (125 miles) from the city of Sydney, Australia. This 40-metre (130-foot) evergreen tree is not strictly a true pine, but a member of the Araucariaceae, a southern hemisphere family that includes kauris, Norfolk Island pine and the Chilean pine or 'monkey puzzle tree'. Before the discovery the genus *Wollemia* and its close relations were known only from fossils in Australia, New Zealand and Antarctica. Some of these date back 90–100 million years, with the most recent at around 2 million years. The tree's specific name *nobilis* was taken not from its 'noble' stature or bearing but from its discoverer, National Parks and Wildlife Officer David Noble.

Experts quickly realized the significance of their find and began a vigorous conservation plan. The small discovery site, with fewer than a hundred mature trees, was kept secret

● WOLLEMIA
Formerly known only from fossils, this tree was unexpectedly discovered in Australia's Blue Mountains in 1994. The tallest specimen is 40 metres (130 ft).

SPOTLIGHT ON... GINKGO

The term 'living fossil' is something of a misnomer, since plants and animals regarded as such are rarely the exact same species as their preserved counterparts from millions of years ago. However, the 40-metre (130-foot) ginkgo comes very close. The single living species *Ginkgo biloba*, also called the maidenhair tree, is highly unusual, with no especially close surviving relatives. Its nearest extant cousins are probably cycads, and then the conifers, making up the flowerless but seed-producing major plant category of gymnosperms, or 'naked seeds'.

Ginkgo fossil history extends back beyond the dinosaurs, more than 260 million years ago (before the 'Great Dying') into the Permian period. Widespread a few million years ago, their range gradually shrank until ginkgos were restricted to eastern Asia. Here they have been planted and cultivated by people for thousands of years, for use in traditional medicine, as timber and as a food source. Since their 'discovery' by Western plant collectors their distinctive and attractive foliage, and fairly hardy habits, have made them welcome guests in gardens, parks and even cities across the world.

A few small populations of ginkgos in China have been regarded as truly 'wild' and listed by the IUCN as 'Endangered'. But this status has recently been questioned. These particular trees have greater genetic similarity than would be expected in natural, wild-sown groups, so they too may have been cultivated.

● GINKGO

The distinctive fan-shaped leaves of the modern ginkgo *G. biloba* (left) are almost indistinguishable from its 60 million-year-old fossilized relative *G. cranei* (top) from North Dakota, USA.

and regularly patrolled. Seed cones were collected and germinated. Trade began in seedlings, first to botanical gardens and trusted private collectors, and more recently to the public, marketed as the 'Dinosaur Tree'. The hope is to establish the Wollemi pine in parks, gardens and other cultivated sites around the world. At the same time the original discovery site will, one hopes, remain mostly intact in its half-million-hectare (million-acre-plus) Wollemi National Park, in the Blue Mountains of New South Wales.

Ultimate survivors – until now

The largest of the seven living species of sea turtles is the leatherback, *Dermochelys coriacea*. This ocean wanderer can reach 3 metres (10 feet) in total length and weigh almost 1 tonne. It is distinct from other sea turtles in that it has leathery skin instead of a bony shell. It can also claim an ancestry that stretches back more than 100 million years, to the Dinosaur Age. In fact turtles and tortoises as a group, the chelonians, date back to the start of the dinosaurs, more than 200 million years ago.

The turtle body plan and lifestyle have changed little over the ages. Why so long-lived? Features include their considerable size, protective shells, ability to go without food for many months, and the large number of eggs they produce. They also travel widely in the oceans. Turtles can swim slowly and steadily for thousands of kilometres in search of their widespread food – leatherbacks eat jellyfish, sea-squirts and similar soft-bodied marine creatures – and to avoid danger. These factors have all been identified as ways in which they have avoided extinction.

Sadly, the turtles' longevity and staying power have not coped with the onslaught of humans. During the Great Age of Discovery, as sailing ships explored and mapped the world's oceans, crews used these enduring reptiles as 'living larder' sources of fresh meat. All seven species of marine turtles are now under threat, with the leatherback, Kemp's ridley and hawksbill listed by the IUCN as 'Critically Endangered'. Despite legal protection, and bans in the trade of live turtles or their body parts, they are still hunted for their meat and decorative shells. Their eggs, laid in beach sand, are harvested for food. And they are tragic regular victims as 'bycatch', trapped in gigantic fishing nets strung through the oceans. These prevent the turtles, as well as dolphins and other air-breathers, from coming to the surface to breathe, and so they drown. As great survivors from ancient times, which have now run into deep trouble, sea turtles symbolize the plight of so much wildlife in the modern world.

● LEATHERBACK TURTLE
Like all marine turtles, the leatherback *Dermochelys coriacea* faces an uphill struggle against human domination.

current crisis
THE SIXTH MASS EXTINCTION?

● WOOLLY RHINOCEROS
Highly adapted to the last Ice Age, this 3-tonne (6,600-lb) beast went extinct in the past 10,000 years. A cause could have been human activity – a trend that increases today.

ARE WE IN THE MIDDLE of another mass extinction? Or at the beginning of one? Regular news reports describe the plights of animals and plants highly threatened for various reasons, and the conservation measures being taken – or not – to try to save them. But is it fair, or even possible, to compare loss of biodiversity and extinctions from recent centuries, or during historical times, or even the past 10–20 millennia, with those that happened over periods of hundreds of thousands of years, and hundreds of millions of years ago?

Comparison calculations run something like this. From the fossil record, taking into account corrections and variables, the typical 'lifespan' of a mammal species through most of time has been about one million years, although this has extended in some cases to 10 million years. Currently there are some 5,500 mammal species. Working from the typical lifespan in the fossil record, we would expect a very broad, average background extinction rate of, say, one species every 100–300 years. History over the past 400 years has seen at least 80 confirmed mammal extinctions, and probably many more of obscure species, and/or in remote places. This is many times the expected background rate. There are many types of calculations like this, some credible, some less so. In 2010 they led the United Nations Environment Programme (UNEP) to consider that Earth is now in the midst of a mass extinction.

Holes can be picked in these types of estimates, as explained later (see also page 93). As with the interlinked subject of climate change, there is also a vast range of conflicting data, and opposed opinions. Some are disbelievers or deniers. They contend that species have always gone extinct, it is all part of a natural cycle, perhaps in the past they did so at today's rates, science is not thorough and clever enough to provide the full story, and there is a lack of knowledge about loss of biodiversity so long ago. Other people shout from the rooftops that the sixth major extinction is well under way. They pronounce that – as with climate change – we have passed the 'tipping point', and we are all doomed.

Yes, no, probably, perhaps

Between the extremes are considered views. Current, and currently increasing, rates of extinction – due largely to human impact – seem to be significantly higher than background rates. As UNEP reported, depending on the animal groups studied – birds, say, or insects – and the time spans, estimates vary from tens through to thousands of times greater. So, if yes, today's extinctions are not just part of a natural cycle, but more serious, should we be taking greater action now? If we wait for so much evidence to accumulate that everyone is convinced, probably we will be too late to help.

How many species?

Scientists can understand any present mass extinction by measuring Earth's biodiversity and the rate at which it has been, is being, and will be lost. This means knowing changes in numbers of species on Earth. But here lies an early challenge. When Carolus Linnaeus developed his biological classification system in the mid-eighteenth century, he listed about 7,700 plant species and more than 4,300 species of animals. In Victorian times, as naturalists and collectors ventured across the globe, these numbers rose rapidly. By the end of the twenty-first century estimates for the total number of species on Earth

SPOTLIGHT ON… IPBES

April 2012 saw the first meeting of the Intergovernmental Platform on Biodiversity and Ecosystem Services, IPBES. (The term 'ecosystem services' is discussed later, see page 74 onwards.) More than 90 nations signed up. IPBES is a 'science–policy interface'. On an ongoing basis it gathers scientific information on biodiversity loss, ecosystem health, habitat degradation, impending extinctions and other complex topics. IPBES includes in this indigenous and local knowledge about conservation and sustainability. This information is packaged, and made available to politicians, leaders and national administrations around the world, to help guide their policies and plans.

Powerful campaigning groups may champion a single narrow cause, perhaps at a cost to others. Also various branches of industry, manufacturing and similar vested interests can be very selective with their facts. IPBES aims to furnish unbiased information that counteracts misinformation and misrepresentation. It plans to perform assessments of biodiversity and ecosystems, and to build databases to aid decisions surrounding sustainable development.

Will IPBES help? Its role is only just beginning. But its climate counterpart, the Intergovernmental Panel on Climate Change (IPCC) – set up in 1988 – is now integral for information-gathering and decision-making on this global, and closely related, topic.

● PREPARING FOR IPBES
Preparatory meetings to establish IPBES began in 2008, with an ad hoc gathering of contributing groups in Malaysia.

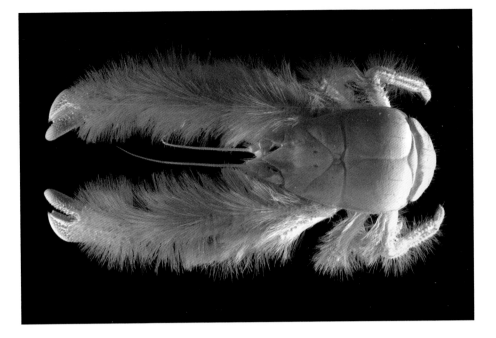

● YETI CRAB

A Census of Marine Life discovery, the scientific name *Kiwa hirsuta* salutes the legendary shaggy-haired 'abominable snowman' of the Himalayas.

● DARTH VADER JELLYFISH

This grape-sized, whimsically-named discovery by the Census of Marine Life – not just a new species but a new genus *Bathykorus bouilloni* – is likened to the *Star Wars* baddie's helmet

were in tens of millions, even as high as 40–50 million. In 2011 a report from UNEP, went lower, to between 7.7 and 10 million, with 8.7 million most likely.

About 2 million species are officially described and named, with two-thirds entered into central databases. The rest – many being insects and other bugs in tropical forests – are unknown to science. Given the rate of habitat loss, especially tropical forests, there are estimates of these undiscovered species going extinct at the rate of 1, or 100, or more per day.

Eighty nations took part in the Census of Marine Life, starting in 2000. By 2010 its workers had formally described more than 1,200 new species from the oceans, with another 5,000 or more in the pipeline of official recognition. During this 10 years of massive worldwide effort, the Census added to biodiversity lists by more than one, and perhaps up to two, species per day.

Feathered elephants

The elephant birds, *Aepyornis*, were massive and flightless, the largest measuring 3 metres (10 feet) in height and perhaps 350 kilogrammes (770 pounds) in weight. They evolved in Madagascar over millions of years. But with waves of human occupation, and especially with European arrivals during the sixteenth century, the birds soon died out. Commonly suggested causes are human hunting for meat and eggs. All elephant birds were gone by the seventeenth century.

The great auk, *Pinguinus impennis*, was a seabird related to other auks such as guillemots and razorbills. It was once widespread around the North Atlantic, nesting on remote rocky islands. As sailing ships became more seaworthy and adventurous from the fourteenth century, the great auk was hunted for its meat and eggs, its oil from being rendered down in industrial vats, and for its exceptionally soft down feathers – greatly prized for pillows. By the 1800s the birds were rare, so museums and collectors raced to pay even more for their feathers, skins and eggs. The last live specimens collected were a great auk pair, with their egg, in 1844 on Eldey Island, Iceland. They were killed. The final scientifically accepted sighting of live birds was from 1852, on the Newfoundland Grand Banks.

Too common to fail?

Before Europeans spread across North America there were at least hundreds of millions, probably billions, of passenger pigeons, *Ectopistes migratorius*. Giant flocks took hours to pass. The pigeons seemed too numerous to be at risk, ever. Then industrial-scale capture and hunting for cheap meat, plus conversion of their oak and beech forests habitats to farmland, decimated their numbers. By the late nineteenth century they were fading fast.

An interesting factor in this decline concerns breakdown in social behaviour patterns. Possibly large gatherings of the pigeons were needed as the natural stimulus for feeding as the birds foraged through the ancient forests, and also for their courtship and breeding. Once the flock numbers fell below a certain level, this instinctive trigger disappeared, and the birds failed to feed efficiently or reproduce. Towards their end, small groups of pigeons were taken into captivity, in attempts to save them. They did not breed, perhaps due to disrupted behaviour. Similar situations are suspected in other gregarious, socially dependent

● **GREAT AUK**
Hunted for meat, oil, eggs and feathers, great auk were extinguished simply because they were so rare.

● **ELEPHANT BIRD EGG**
One of these eggs, more than 30 cm (12 in) long, provided the equivalent of 160 hen eggs – a sizable omelette!

species, such as the slender-billed curlew described later (see page 49). In the end, probably the world's last passenger pigeon, the celebrated 'Martha' of Cincinnati Zoo, passed away in September 1914.

Sloane's urania, *Urania sloanus*, was a spectacular moth from Jamaica. Even before Europeans arrived, its numbers seemed to boom and crash. Anecdotes describe moths breeding so well when conditions suited the hungry caterpillars or larvae, that they ate out the local food supply – mainly *Omphalea lianas* of the spurge family. So the adults migrated to a fresh area of lowland forest. This boom-crash-migrate cycle required large areas of forest. When Europeans began clearance for sugar cane and other crops, the urania waned. The last moth specimens were reliably known from the mid-1890s, about 170 years after the species was described for science.

One of the 'speediest' extinctions, in terms of the gap between formal description and last sighting, was Steller's sea cow, *Hydrodamalis gigas*. This enormous herbivore was a member of the sirenian mammal group, which contains four living species – three manatees and the dugong. They are big, but much smaller than the mammoth Steller's sea cow of the far north. It reached 9 metres (30 feet) in length and weighed up to 10 tonnes.

● SLOANE'S URANIA
Being a day-flier helped collectors to net this beautiful moth for museums and admirers as it became rarer.

● PASSENGER PIGEON
These birds may have needed the social cues of vast flocks and colonial breeding to reproduce successfully.

Steller's sea cow was described by a sailing expedition naturalist Georg Wilhelm Steller in 1741. Around the North Pacific Rim, from Russia to California, prehistoric people had already hunted to oblivion this slow, peaceful creature, for its meat and durable hides. Only a few small groups remained in isolated Pacific islands far away from mainland. Following Steller's exposé back in Russia, the last remnant populations were rapidly located and finished off by Russian fur-hunting expeditions, which regularly stopped over to stock up on provisions over the next two decades. The giant was probably extinct by 1768 – less than 30 years after its published discovery. Its scientific name, *Hydrodamalis gigas*, was published in 1780, 'post mortem'.

● SEA COW
Steller's sea cow was three times longer and 15 times heavier than its close living relation the dugong, *Dugong dugon*, shown here – itself a threatened species.

Disrupted food chains

One of the most high-profile recent extinctions is the baiji, also known as the Chinese river dolphin or Yangtze (Chang Jiang) River dolphin. At least, it is presumed gone. This dolphin endured years of decline, from perhaps many thousands in ancient times to a few hundred in the 1980s. Despite a ban on hunting, its demise continued to less than

The baiji *Lipotes vexillifer* could reach 2.5 metres long, 150–200 kilograms in weight, and was bluish grey or grey above and ashy white below. Its habitat of cloudy, muddy waters meant vision was little use, so the baiji had tiny eyes. Instead, to navigate, forage for prey and avoid predators, this dolphin relied on a form of the sonar employed by submarines. It made clicks and other sounds and listened for echoes – the same system used by bats, termed echolocation.

The baiji faced a legion of threats in the rapidly industrializing Yangtze and its tributaries. Noise pollution interfered with its sonar. Chemical pollution may have weakened and poisoned its body.

● BAIJI

The baiji's name *Lipotes* comes from the Greek 'left behind', which it has been in that it has probably gone.

Injuries from boat hulls and propellers were common. It became tangled in fishing nets and lines and drowned, and was stunned to death by electro-fishing. River water levels fluctuated greatly as water was taken for farmland, livestock, human use and industry. Huge irrigation and hydro-electricity dams, such as the giant Three Gorges Installation, blocked its movements and those of its prey, changed river and sediment flow, and contributed to the further corruption of an already much-degraded habitat. Sadly, no contest.

100 by the mid-1990s. The last confirmed sighting was in 2004. A thorough search of the baiji's likely remaining range in 2006 failed to find any. In 2007 most scientists accepted that the baiji had gone, with the IUCN considering it 'Critically Endangered' until extinction criteria are met. Even if a few baiji miraculously survive, they would probably be too small in number, with too limited a gene range, to re-establish a healthy population – a situation dubbed 'functional extinction'.

These plant-eaters probably disappeared because of alteration to the limited habitats where they laid eggs.

Not just pretty ones

Past mass extinction events have affected most forms of life, not just the cute or cuddly. Recent extinctions are similar, affecting not only creatures who generate public interest and sympathy but also those who are disliked or even reviled. In North America the Rocky Mountain locust, *Melanoplus spretus*, once swarmed in billions. One gathering was perhaps the greatest known single concentration of individuals from one species in one place – an estimated 12,000 million! As natural prairie and scrub went to crops and grazing in the early nineteenth century, the locusts revelled in plentiful food. Then after outbreaks in the 1870s, its numbers fell dramatically. It is thought that habitat destruction caused by agriculture in the river valleys, where the locusts laid their eggs, was to blame. This destroyed the first stage in the insect's life cycle. The last Rocky Mountain locusts were seen in 1902. It was another turnaround of 'too many to fail'.

Snails may not be everyone's choice for creatures to conserve. But they are vital components in most ecosystems, as common consumers of plants, small prey or scavenged material, who in turn provide food for all manner of predators from shrews and birds to badgers and wild dogs. The giant African land snail, *Achatina fulica*, is a big-eating, fast-breeding species. Originally from East Africa, it has been introduced to many regions around the world, both as food and as an exotic 'pet'. As usually happens in these instances, the snails escaped – and were released – into the wild. They now cause havoc as they out-compete local snails and other creatures for food and living space. They affect crops and also spread diseases. As a consequence, the giant African land snail is now listed in the world's top 100 invasive species (see page 55). Pest eradication programmes with molluscicide poison chemicals, hand-collecting and even flame-throwers have had only limited success.

● O'AHU HAWAIIAN TREE SNAILS
Many colourful variations of these species are extinct due to the Florida rosy wolfsnail.

Changing tastes

The giant African land snail is certainly not at risk of extinction. But its introduction has led to several. These snails were brought to Hawaii on two known occasions in 1936, once in mail from Japan and once in luggage from Taiwan. The snails escaped and began to damage farm crops and local nature. After trying various methods of eradicating this invasive species there followed an attempt at biological control – using natural enemies to limit another species, often itself introduced, such as by eating it, usurping its living space, or transferring disease.

Another snail was flown in: the Florida rosy wolfsnail, *Euglandina rosea*. Also known as the cannibal snail, this is a flesh-eater with a penchant for other snails. However, the cannibal snail found native Hawaiian snails, especially O'ahu tree snails, *Achatinella*, more to its liking. Within a couple of years several *Achatinella* species were extinct, devoured by the newcomers. Of the 39 species of O'ahu snails, prized in Hawaiian traditions for their colourful patterns, 15 are listed by the IUCN as 'Extinct' and the remaining 24 as 'Critically Endangered' (although several of these have probably now also gone).

● BUG SURVEY
Newly discovered species of insects regularly add to our biodiversity databases, especially from tropical rainforests. The tent-like malaise trap guides flying insects into a collecting pot at the top.

Both the giant African land snail and the cannibal snail continue as serious pests in Hawaii and on other Pacific islands. In effect, two seriously destructive invasive species for the price of one, with dozens of extinctions thrown in free.

How do we know?

How can we be sure of the status for extinct or nearly-so species, from dolphins and birds to locusts and tree snails? How do we measure biodiversity, and identify, count and assess plants, animals and species? At the 'sharp end' in the field – out in nature – biologists, naturalists, rangers and others have a wide array of methods to obtain the most accurate and reliable information.

Field surveys take many forms. Highly trained staff with specialist equipment can undertake a dedicated survey targeting one region and perhaps only selected species. Local people come on side to help with their casual observations from everyday life. All of these might look for the animals or plants themselves, also their remains such as shells, bones or plant cones, and signs such as animal droppings, nests, scratchings or feeding marks on vegetation. Various forms of sampling look for pollen in the air or tiny eggs in water.

Advances in automatic equipment, such as cameras triggered by movement or infrared (heat), have helped enormously. These can be left for days, even weeks, and are especially helpful in finding nocturnal or night-active animals – from those believed to be extinct, to species new to science. More traditional methods are nets and traps. These are specially designed to hold creatures temporarily without harm. Such methods give scientists the opportunity to release unwanted catches and ensure that any essential collecting is more selective and targeted.

Lost worlds found

In 2008–09 a BBC-based team of more than 40 zoologists, other scientists, and documentary-makers spent months in Papua New Guinea, Southeast Asia. They were first to carry out a scientific survey of Mount Bosavi, the 4-kilometre (2½-mile) wide, isolated crater of a long-inactive volcano. Steep ridges around the crater make this a 'lost world' rarely visited by any humans, where plants and animals evolve in isolation. Here the expedition discovered over 40 new species and subspecies, including insects, spiders, fish, frogs, a bat, a tree kangaroo and a rat. The Bosavi woolly rat, a species of the genus *Mallomys*, is one of the world's largest rats, 80 centimetres (more than 30 inches) long and weighing 1.5 kilogrammes (3¼ pounds). It had no fear of humans, having not experienced them as a threat.

Another battery of methods is used to analyse and interpret information from the field. This is the 'armchair' (more usually office bench) end of the operation. One vital task is to give a firm identification to the specimens found – from a new species, to one believed extinct. For this, museum collections and taxonomy databases are invaluable.

Patterns and concepts

There are millions of species, and it is impossible to measure every one. Scientists may try to focus efforts, for example on a keystone species – one that has lots of influence over its ecosystem, even if it is not especially numerous. This limited information is collated and unravelled, and patterns may then emerge. Changes in local biodiversity can indicate what is happening regionally, or even across the world, to unstudied species. In this way global trends are extrapolated from small samples.

The keystone species has a pivotal role in its ecosystem that may seem out of proportion to its presence, and this is an important concept in conservation biology. For example, a small and innocuous (to us) plant-eating bug may keep in check its main food species. It could do this by nibbling the plant's germinating seeds, so very few such bugs are able to have an outsized effect on keeping the plants in check. The relationship is in balance after a long evolution. But if the bug is reduced, such as by chemical sprays or introduced predators, its plants can quickly multiply. The plant bloom may alter the character and balance of the whole living system, even threatening other species with extinction.

In addition to keystone species, scientists use other concepts to analyse ecology and biodiversity – although some are partly overlapping and not always defined in a watertight way. For instance, indicator species can be thought of as plants or animals that are especially sensitive to health or disturbance in an ecosystem and are useful to monitor for early warning signs of problems. Ecosystem engineers are species that dramatically change their habitats or even create new ones, such as beavers damming a stream to create a pool. On the North American plains, once-vast prairie dog 'towns' altered the environment over huge areas, dictated which plants could grow where, and provided underground sites for nesting birds such as burrowing owls, and also retreats for snakes, lizards and others.

So many threats

Here's a 'short'-list of threats to biodiversity due to human activities, with categories by no means mutually exclusive:

- habitat destruction, the over-arching threat, for so many reasons – farming, industry, human settlements, timber, mineral extraction, transport (and golf courses)
- hunting, harvesting and general over-exploitation for food (including bushmeat), trophies, 'sport' and more
- warfare and military struggles, which disrupt local habitats in so many ways
- pollution of air, soil or water, or all of these
- increased urbanization and industrialization
- intensive fishing and farming
- introduced and invasive species competing with, predating on, and generally ousting local ones

- spread of diseases, such as from introduced domestic plants or livestock to related native ones
- emission of GHGs (greenhouse gases), leading to the vast global challenges of climate change and global warming.

How causes combine

Causes of threat and extinction interact in complex, often unanticipated ways. The slender-billed curlew, *Numenius tenuirostris*, is a mysterious wading bird that breeds in the far north, on the briefly thawed bogs, marshes and other wetlands of the taiga – the vast conifer forests of northern Asia. In autumn these long-legged migrants come to the Mediterranean and surrounding regions to overwinter on shallow freshwater marshes and lagoons, where they pick-and-peck for small worms, bugs, shrimps and similar food.

The curlews' numbers have been in decline for some time, and in the past 20 years sightings have been even more infrequent, with just one or a few birds each time. Why? The breeding areas are vast and so far little modified, in relative terms, by human activity. The stopover habitats on migration to rest and refuel, and the wintering quarters, include wetlands such as coastal lagoons, saltmarsh, shallow lakes, inland swamps and lake margins. Most of these habitats are in steep decline, drained for agriculture, livestock grazing, industry, human habitation and tourism. Also, on their wintering grounds, slender-billed curlews and many similar birds have long been hunted by shooting, trapping and other means. This last activity is probably a major factor in the species' long-term demise.

Another factor concerns the birds' social behaviour patterns. It is, or was, a gregarious species, migrating and wintering in flocks. Perhaps the birds cannot function effectively when they are just a few. For example, they may need the stimulus of many pairs of whirring wings, or lots of calling individuals, to trigger the instinctive action of taking to the air, to find a fresh feeding site. Once numbers become too low, this behaviour pattern can no longer function – as for the passenger pigeon, mentioned earlier (see page 42). Hunting and habitat loss may have combined with failed social behaviour, due to a fall below critical numbers, into a vanished chance of recovery.

● SLENDER-BILLED CURLEW
Sightings are vanishingly rare and, due to similarity to related species, highly disputed. A single bird in Hungary in 2001 is one fully accepted and internationally verified record from the 2000s.

SPOTLIGHT ON... FIJIAN GIANT LONGHORN BEETLE

The island nation of Fiji, in the southwest Pacific, is home to three of the world's largest insect species. Enormous beetles of the genus *Xixuthrus* are collectively known as Fijian giant longhorns. From front of head to end of body, the largest measure 15 centimetres (6 inches), with the long, powerful legs adding perhaps half as much again. These mega-bugs are all rare and, as a species group, under severe threat. Their wood-munching larvae depend for food on huge hardwood trees in good-quality lowland rainforest. But such massive tree specimens are selectively logged because their wood is economical to process, mature and valuable. Also much lowland rainforest has been cleared for agriculture. Added hazards are powerful cyclones that sporadically flatten areas of forest, and the beetles' very slow reproduction rate – they remain as larvae for 12 years or more. The IUCN has not yet evaluated *Xixuthrus*, but the future of these spectacular creatures must be, to say the least, uncertain.

● GIANT FIJIAN LONGHORN
The 'horns' of these beetles – here a male *Xixuthrus terribilis* – are in fact their feelers or antennae.

The slender-billed curlew is the subject of several conservation initiatives. Scientists have analysed its feathers, droppings and museum skins for chemical substances termed isotopes. These vary in soil, water and living things from region to region, and give clues to where the birds breed, stopover and overwinter. But sightings of the slender-billed curlew have almost ceased. The IUCN lists the species as 'Critically Endangered' and there may be fewer than 50 adult birds left.

Limits of flight

As described later (see page 57), islands are both 'hotspots of evolution' and suffer greatly from extinctions. Two very different birds illustrate some of the common dangers faced in these limited habitats, as well as hazards unique to each species.

The Alaotra grebe, *Tachybaptus rufolavatus*, was limited to Lake Alaotra, the largest lake in Madagascar, and a few nearby locations in the country's north. Small wings meant its flight powers were seriously limited. Most of the grebes lived sedentary lives on the main lake. But changes here have been drastic. As surrounding agriculture increased, run-off pollution and sediments lowered the lake's water quality. Introduced plants crowded out natural vegetation, and invasive fish disrupted food chains. Some of these introduced fish preyed on the grebes and their young, such as the 1 metre (3 1/3 feet) snakehead murrel, *Channa striata*, a fierce carnivore from southern and Southeast Asia. Increasing use of nylon gill-nets and fishing lines has affected many of the lake's waterbirds, and the birds are also caught locally for food.

The last confirmed sighting of the Alaotra grebe on its home lake was in 1982. Further 'sightings' in the mid-1980s may have been misidentified, or of birds interbred with other grebe species. Checks in 1999 and 2000 failed to detect any Alaotra grebes. Given its lack of mobility, and few other suitable habitats in the region, in 2010 the IUCN declared the grebe 'Extinct'. In effect, it lived within two 'islands' – one its lake surrounded by land, the other Madagascar itself. There is just one known photograph of the species.

Plight of the palila

The Hawaiian honeycreeper known as the palila, *Loxioides bailleui*, has well-known ecology, and its threats are much studied. The palila uses its stout, finch-style bill to feed on immature seeds of the mamane, a native shrub or tree from the legume (pea and bean) family. In season, these are virtually the bird's only food, although it switches to seeds, fruits and berries at other times. The palila also eats larvae or caterpillars of codling moths, *Cydia*, which themselves feed on mamane seeds, and other moth larvae that use the mamane tree, such as snout moths, *Uresiphita*. These food sources are important as a source of protein to the adult birds, and even more so to their hungry nestlings. In fact even the palila's nests are in mamane branches. The bird is highly dependent on the tree.

The palila is limited to mamane areas of dry forests at altitudes of 2,000–3,000 metres (about 6,500–10,000 feet), on the slopes of the famous volcanic peak Mauna Kea. The species' entire world population of fewer than 1,000, and declining, is limited

● ALAOTRA GREBE

These Natural History Museum specimens from 1929 were used to define the species in 1932, helping to identify other individuals as pure-bred, or hybrids with closely related grebes.

● PALILA
Human interference and introduced species, from sheep to predatory ants and parasitic wasps, menace the already miniscule range of this honeycreeper on Mauna Kea, Hawaii.

to about 200 square kilometres (77 square miles) here. Currently listed as IUCN 'Critically Endangered', the palila might not even survive today if not for earlier action. In 1967 it was put onto the US Endangered Species Act list. In 1978 a legal challenge by conservationists led to feral goats and sheep being removed from some of the mamane forests. But there are many other dangers: disruption of the area by off-road vehicles, military training and human-started fires; predation of eggs and nestlings by black rats and domestic cats; invasive plants that stifle mamane early growth; even introduced wasps and other insects preying on or parasitizing the caterpillars so important to the palila's breeding.

Climate change and coral

The multitudinous effects of climate change and global warming on biodiversity, and the demise and extinction of species, cannot be overstated. It is another pressure to add to habitat destruction, pollution and the rest. Across the globe, climate patterns are predicted to disrupt and dislocate. Seasons are moving in place and time. Rains fail in traditionally wet regions, while downpours deluge dry zones.

Climate change has its doubters and deniers. They claim either that it is not happening, or that it is part of a natural cycle and humans are not to blame. Many scientists take the view, considering present evidence, that climate change is here, the planet is warming, and there is a 90 per cent or more certainty that greenhouse gases produced by human activities are the chief cause.

In all tropical oceans, coral reefs – among the most biodiverse ecosystems on the planet – are dying. World Wide Fund for Nature, WWF, consider that one-quarter of all reefs are damaged beyond repair, and another two-thirds are seriously threatened. As well as climate change, many other factors are involved: destructive fishing with cyanide, dynamite or bottom-trawl nets; general overfishing; poor tourism practices such as uncollected beach litter and untreated effluent; pollution by industrial wastes, oil and agrochemicals; land erosion leading to sedimentation; the increasingly acidic character of the water; and coral mining.

More and more coral reef life-forms are at risk – from the coral creatures themselves to sea-slugs, urchins, crabs and shrimps, sharks and other fish, turtles and sea-snakes, and seabirds. If the water temperature rises too fast the tiny aneomone-like coral animals, called polyps, cannot survive. This is because the micro-plants called algae, which live in helpful partnership within the polyps, are expelled. The polyps become colourless and moribund, since they depend on the algae. Gradually reef turns lifeless, pale and 'bleached'.

Bleaching also occurs as the ocean's waters become more acidic. Rising carbon dioxide levels in the atmosphere (a major cause of the greenhouse effect and global warming) mean that more of the gas dissolves in seas and oceans. This forms more carbonic acid and other acidic substances than are usually present, a process is known as ocean acidification.

● CORAL BLEACHING
The central corals here still survive, but the pale listless appearance of the growths around indicate the reef is in deep trouble.

SPOTLIGHT ON... TIMOR REEF-SNAKE

The dusky or Timor reef-snake, *Aipysurus fuscus*, lives its entire life in the ocean, catching fish and other prey with its highly venomous bite, and even giving birth to live young at sea rather than returning to land to lay eggs. The Timor reef-snake is one of the lesser-known of the 60 or so sea-snake species. It is recorded from a handful of shallow reefs in the Timor Sea, between northwest Australia and Indonesia, the reefs covering an area of only 500 square kilometres (less than 200 square miles).

From being quite common at some sites, estimated numbers have plunged by more than 70 per cent since the late 1990s. Yet one of its main reefs has been a protected marine reserve since the early 1980s, and the species is not especially targeted by fishing or trapped as bycatch, as are some other marine snakes. Current thoughts are that a rise in water temperature is affecting the whole coral ecosystem, and it may even be too high

● TIMOR REEF-SNAKE
About 60 cm (2 ft) long, this sea-snake feeds mainly on fish, especially wrasses and gobies.

for the snake itself to withstand comfortably. (Another species, the yellow-bellied sea-snake, *Pelamis platura*, is known to suffer greatly and even die if the water is warmer than 35°C or 95°F.) IUCN considers the Timor reef-snake 'Endangered'.

The shock of the new

Animals and plants naturally venture into new areas, perhaps adapting to local conditions as they go. They may survive and come into balance with local species of predators, prey, competitors, rivals, parasites, diseases and so on. It tends to be a gradual process, taking many thousands of years.

Humans translocate species across continents and oceans in an eye blink – just a few hours with modern air travel. Some of these introductions are intentional, such as exotic plants to brighten up parks and gardens, or fish and waterbirds for ornamental ponds and lakes. Others are accidental, as when specialized pets like snakes and spiders manage to escape into the wild.

It may happen that an introduced – non-indigenous, non-native or alien – species does not survive. Its new location is too hot, too cold, lacks suitable food, or has another challenge. But there are thousands of examples of introduced species soon becoming too successful. Freed from the natural checks of their homeland – predators, competitors, disease, harsh seasons, lack of breeding sites – they multiply with hardly any restraint. The introduced species has become an invasive one, and now it may well threaten some locals with extinction.

Adapt and survive

Invasive species tend to have a suite of features or traits that help them run riot in their adopted home. They usually grow fast, mature rapidly, produce plenty of offspring, spread quickly, are adaptable in growing conditions or feeding preferences, have a wide habitat tolerance – and, most importantly, associate with humans. They out-compete rival natives by being more efficient at gathering nourishment – for example, an introduced predator can devastate populations of prey who are not familiar with its hunting techniques.

Animals infamous from this category include the domestic cat, house and field mice, brown and black rats, grey squirrel, common rabbit, goat, red fox, mink, stoat, Indian mongoose and pigeon, as well as plants such as the kudzu (Japanese vine) and water hyacinth. The IUCN's list of the 100 worst invasive species, maintained on the Global Invasive Species Database, also includes the common starling, Nile perch, red-eared slider terrapin, cane toad, giant African land snail, common prickly pear cactus, yellow Himalayan raspberry. Several of these invaders are mentioned many times in the book. (Oddly for some, *Homo sapiens* is not on the Worst 100 list.)

The small wader known as Saint Helena plover, *Charadrius sanctaehelenae*, from the remote Atlantic island of that name, has been overwhelmed by introduced cats and rats. Also its eggs are eaten by another of the 'Worst 100', the common mynah bird. This plover is ranked by the IUCN as 'Critically Endangered', with just a few hundred remaining. Further threats include potential developments on the island such as a possible airport and likely associated increase in tourism.

Invaders down under

Australia has endured many extinctions at the hands of invasives. The red fox and cat are serious predators of small native wildlife, while the common rabbit's grazing has destroyed millions of hectares of natural vegetation, with consequent knock-on effects on the herbivores and so the carnivores. The lesser bilby or lesser rabbit-eared bandicoot, *Macrotis leucura*, was a small omnivore of the continent's central regions. A marsupial (pouched mammal), its long ears and large rear legs looked uncannily rabbit-like, although it had a more pointed face and a long tail. Its diet was largely bugs and small vertebrates, although it also consumed vegetable matter. Faced with degraded habitats and food competition courtesy of the rabbit, as well as predation from foxes and cats, it disappeared from view by the 1960s and is now regarded by the IUCN as 'Extinct'.

Another victim of the rabbit was the Laysan crake or Laysan rail, *Porzana palmeri*. This small, neat, flightless bird inhabited the island of Laysan, Hawaii. When rabbits were

● ST HELENA PLOVER
The island's national bird is featured on the island's flag and coat-of-arms. One conservation aid is nest 'exclosures' designed to keep out cats.

LESSER BILBY

The last specimens collected alive were in 1931 near Cooncherie, north-eastern South Australia.

introduced in the 1890s, along with hares and guinea pigs, they destroyed the native vegetation and turned much of the island into sandy waste. The last crakes were seen in the 1920s. Meanwhile some had been taken to Midway Atoll, across the Pacific to the west, but they fared little better. With the fierce military conflict known as the Battle of Midway in the Second World War, the Laysan crake was probably terminated by 1944. It is IUCN-listed as 'Extinct'.

Perhaps the most ironic threats and extinctions are due to introductions of new species as biological control agents, often in attempts to check other invasive species that have already run wild. Again Australia furnishes many examples. Perhaps most infamous is the sizeable, powerful, highly toxic cane or marine toad, *Rhinella (Bufo) marinus*. Its story in Australia began with another introduced species – sugar cane, cultivated in enormous plantations in the northeast of the continent. Local beetles soon discovered this marvellous new feast, multiplied and spread.

To eradicate these pests, in 1935 cane toads were imported from their native South America to eat the beetles. The toads thrived, sadly not by feeding on the beetles, but on large portions of the native wildlife. They spread fast and now threaten many animals, including northern quolls, *Dasyurus hallucatus*. These native cat-like marsupial predators, unaccustomed to the newcomers, try to eat the poisonous toads and seem to be especially sensitive to the toxins produced by its skin glands. As cane toads continue to invade new areas, quoll populations crash – not helped by the usual pressures of habitat loss and predation by, or competition from, other introduced species. The northern quoll is already extinct in parts of its former range, and regarded by the IUCN as 'Endangered'.

LAYSAN CRAKE

Only 15 cm (6 in) long, this crake often bullied larger birds away from food such as seabird eggs, but such aggression could not save the species.

● QUOLL
Known as Australia's 'marsupial
cats', quolls are declining in many
areas, partly due to competition for
prey from foxes and cats, as well as
cane toad poisoning.

From hotspot to death row

More than half of confirmed extinctions in the past 400 years have been on islands. These places are well known as biodiversity hotspots. Each island has its own local conditions, habitats and environments. Land animals or plants newly arrived from elsewhere may have the chance to adapt and thrive. As they do so, isolated from others of their kind, they become new species. Then more species may arrive. These might fail to gain a foothold or integrate – or they could turn into the next big-time invaders. This is what humans and their co-travelling animals and plants do. Established native species often have specialized habitat needs, a limited range, and nowhere else to go except the open ocean. They are easily destroyed. Examples are littered throughout this book.

The vast majority of islands around the world, from tiny Saint Helena and Mauritius, to groups like Fiji and Hawaii, to larger New Zealand and Madagascar, have suffered hundreds of extinctions. Currently thousands of island species are under threat, and the trend is inexorably upward. The situation is similar for aquatic species in rivers, lakes or inland seas, which are 'islands of water' surrounded by inhospitable land.

Symbol of extinction

Perhaps the most famous extinction of all time, byword and icon for the fate, is the dodo. About 1 metre (3 1/3 feet) tall, weighing 15 kilogrammes (33 pounds), with a huge beak and tiny wings, this flightless cousin of pigeons and doves had an easy life on Mauritius. It is first mentioned in Western historical records around 1599. Docile, since it had no reason to fear humans, it was a boon for passing hungry sailors. However, more serious threats were probably predation or competition from introduced species, being the usual cats, dogs and rats, and also pigs and crab-eating macaque monkeys (both of these being members of the 'Worst 100' club). The last reliable dodo records date from the 1660s, although it may have clung on in isolated coastal regions until later that century.

This moa skeleton, about 5,000 years old, shows the bird's heavy body, tiny wings and long, powerful hind legs.

Flightless birds

Like elephant birds on Madagascar, moa – genus *Dinornis, Pachyornis* and others – were flightless island inhabitants, in this case in New Zealand. There were probably 10 species, the largest being the giant moa, *Dinornis*, some 3.5 metres tall and weighing over 200 kilograms. Mammals, and especially large mammal herbivores, never got going in New Zealand. Their niches were taken by birds, including moas which ate all kinds of plant matter, from leaves and buds to seeds and twigs.

Humans arrived in New Zealand probably in the late 1200s; exploring Polynesian settlers who became the Maori. Close scientific scrutiny of the most recent preserved moa remains – feathers, skin, muscles – suggests most, if not all of these birds had gone by about one century later. Occasional moa sightings endured into the 1700s but none is backed up by hard evidence. Catastrophic hunting of the birds, also egg collection and habitat change, were probably the primary factors.

Only one predator could tackle a large moa – Haast's eagle, *Harpagornis moorei*. It was New Zealand's equivalent of a tiger with wings. Possibly the biggest eagle of all time, it had a three-metre wingspan and weighed up to 15 kilograms. Despite isolated sightings over the years since, probably by 1400, with its natural prey gone, this spectacular aerial predator also went extinct. The demise of Haast's eagle may have even been hastened by direct persecution from the new human settlers. The bird was big enough to regard smaller individuals – perhaps toddlers at play – as prey.

● HAAST'S EAGLE
The great airborne hunter would have attacked and killed part-grown moa, to eat on the spot.

Mounting evidence

What can we learn from this long list of tragic losses, and hundreds more like them? Biodiversity is reducing rapidly. Extinctions are occurring all around us. Each disappearance is different, with its own sad story to tell. Causes combine. Nature is full of surprises and untold intricate detail for each animal or plant's life history, ecology and survival.

In 2012 the IUCN released its latest set of Red List data, the main official account of what is in trouble where in the natural world. As described later, it contains not only more species assessed, but proportionally more threatened species in every category. If there is an impending sixth mass extinction, it could happen much faster than previous great extinction events. Could it include the species responsible – ourselves?

● ST HELENA TROCHETIOPSIS
Of tiny St Helena's three *Trochetiopsis* shrub species the blackwood is 'Extinct', the redwood 'Extinct in the Wild', and the ebony 'Critically Endangered'.

self-destruction
ARE HUMANS HEADING FOR EXTINCTION?

ANIMALS AND PLANTS OF THE LIVING WORLD are categorized into basic breeding-block units known as species. All humans on the planet belong to one species, *Homo sapiens*, ironically 'wise person'. In evolutionary terms we are just an eye blink, having arisen in our current physical form less than one-fifth of a million years ago. Also physically, as a member of the mammal group, we are quite ordinary – far less extreme or remarkable, in terms of special adaptations and extravagant features, than the blue whale, giant elk or sabre-tooth cat. Looking back at the immense time span of Earth, with its background extinction rates, we might be expected to last another 1, 2, or maybe 5 million years.

Modern humans have changed the world so much since we have been here. Less than a century ago, our mental abilities led to the invention of nuclear weapons. At one time enough of these devices were active to annihilate life across most of the planet in just a few hours. However, reason and perhaps species self-preservation prevailed. Rival nations retreated from the brink of self-destruction. Some argue that we have now chosen a longer, slower path towards the same end, with major ecological changes such as acid rain, ozone depletion, unsustainable use of natural resources, a rising energy crisis, many kinds of pollution, climate change and global warming... Sooner rather than later, as well as much wildlife, we too may be gone – perhaps renamed, by any successors, as *Homo extinctus*.

The road to modern humans

We are by no means the first of our group. Several other humans, genus *Homo*, and many close relatives have been and gone in the past few million years. All were members of the major mammal group the primates. Fossils show this appeared more than 50 million years ago, and from evidence of molecular studies, perhaps even more than 80 million years ago – before the non-bird dinosaurs went extinct. Early primates resembled lemurs of today, which are one main branch of the group.

● SAHELANTHROPUS
Toumai ('Hope of life') from Chad is the most complete specimen of *Sahelanthropus* – a close relative or even ancestor of humans and chimps.

Around 30 million years ago along came monkeys, then the apes, hominoids. The gibbon or 'lesser ape' line of evolution split from 'great' apes by 15 million years ago. The line leading to today's only living Asian great ape, the orang utan, separated at 14 to 12 million years ago. African great apes, hominids, divided into the gorilla and the chimp–human branches around 10 million years ago. By 7 to 6 million years ago chimps had separated, and the evolution towards ourselves was under way.

The fossil record continues to give clues to our extinct relatives and direct ancestors. Most of these were in Africa, but all except ourselves have gone extinct. The human past is hotly debated as to who evolved into what, and who was a separate species or not. After the split from chimps, upright-walking *Orrorin* possibly lived in what is now Kenya, some 6 million years ago. *Ardipithecus* dates from more than 5 to 4.4 million years ago and has a mix of chimp and human features. Various kinds of *Australopithecus* evolved and spread in Africa from 4 to 2 million years ago, including the famous upright-walker 'Lucy'. Some of these became heavy in build with very powerful jaws, and they may be a separate genus *Paranthropus*.

● AUSTRALOPITHECUS
Australopithecus afarensis, the species to which 'Lucy' belonged, lived in northeast and East Africa.

● PARANTHROPUS
These early humans, shown here by 1.8 million-year-old 'Nutcracker Man', had large, powerful jaws and teeth.

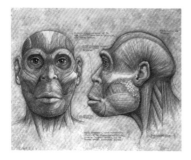

Homo gains and losses

The first humans were probably the slight, smallish-brained *Homo habilis*, usually taken to mean 'handy human', 2.3 to 1.3 million years ago. Evidence of tool-using dates from this time. By 1.8 million years ago 'work human', *Homo ergaster*, appeared, although it may have been the same species as 'upright human', *Homo erectus*. The latter was now looking much like modern humans, and strode not only in Africa but also migrated to Asia. Two more species/subspecies were 'pioneer human', *Homo antecessor*, 1.2 million to 800,000 years ago in Europe, and 'Heidelberg human', *Homo heidelbergensis*, younger at 600,000–400,000 years ago, with a larger brain and better stone tools, in Europe and Africa. This gallop through human evolution mentions only selected species. Some experts suggest fewer than 15, others go for more than 25. It is one of science's most hotly contested topics. But all have gone.

The well-known Neanderthal people, *Homo (sapiens) neanderthalensis*, battled the last great ice age in Europe. They survived from more than 350,000 until 30,000 years ago, possibly 25,000. By this time our own species *Homo sapiens*, having arisen in Africa perhaps 200,000 years ago, had also arrived in Europe. Genetic and some fossil evidence suggests that the Neanderthals may have interbred with modern humans, perhaps some time between 80,000 and 40,000 years ago in the Middle East. In 2010 the Neanderthal Genome Project estimated that modern humans of non-African descent carry 1 to 4 per cent of Neanderthal genes among their own.

Why did Neanderthals go extinct? Two main causes are suggested: climate change and an invasive species, namely ourselves. (Sounds familiar?) Neanderthals were stocky and short-limbed, 'cold-adapted' to lose less body heat in cool conditions. The climate was changing rapidly from 60,000 years ago, and plants, animals and whole habitats shifted and evolved. Then, from 45,000 years ago came the newcomers with their sophisticated tools and cultures. They possibly out-competed the Neanderthals for food and other resources, spread diseases to which the Neanderthals had less resistance, even came into direct conflict, and/or interbred them to oblivion.

● SKULL, GREECE
This skull is possibly *Homo heidelbergensis*, perhaps the ancestor of Neanderthals and ourselves.

Most recently extinct

Neanderthals may not have been the most recent humans to go extinct. The Flores 'hobbit people' are one possible contender. Another is the Red Deer Cave people. Their fossils from southern China, first found in 1979, date to only 11,500 years old, and show a strange mix of ancient and modern features. Further studies may class them as a distinct species.

● NEANDERTHAL SKULLCAP
The first Neanderthal fossil was a 40,000-year-old 'skullcap' found in 1856 in a Neander Valley cave, Germany.

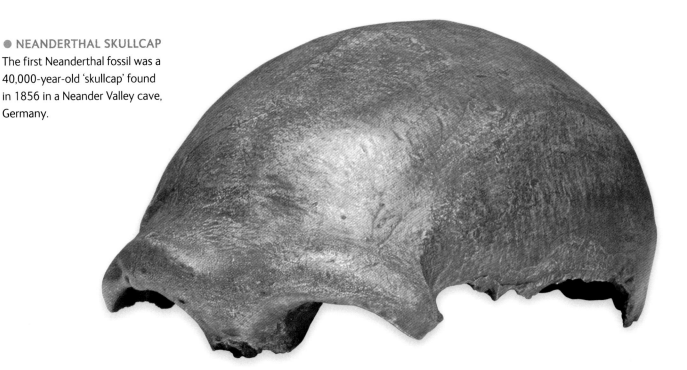

The most recent human extinction may be of the so-called 'hobbit people', provisionally named *Homo floresiensis*. Their remains were discovered on the Indonesian island of Flores in 2003 and date from at least 40,000 to as recently as 13,000 years ago. Among them are parts of up to 10 individuals plus stone tools, and, in particular, a female skull. The evidence suggests a diminutive human about 1 metre (3⅓ feet) tall with a very small head and brain. The bones show a curious mix of features, some relatively modern, others harking back to long-gone humans such as *Homo erectus*. The tools are also puzzling, being quite sophisticated and of a type dating back to 90,000 years ago.

These remains are subject to intense scientific studies, including chemical analysis, anatomical examination, X-rays and scans. The results are clouded by an unseemly collection of vested interests (politicians, scientists and others) who guard specimens and results jealously – what is known as 'hobbit politics' – as they strive to win by pinning down the result.

The 'hobbits' may have evolved as a pygmy group of islanders, or as a distinct species, an example of the well-known phenomenon of

● 'HOBBIT' HUMAN
The original tiny skull and jaw of *Homo floresiensis*, with a lower jaw from a different individual of similar size.

island dwarfism. Or they could have been a line of modern humans affected by a medical condition such as microcephaly, or a genetic anomaly such as an inherited form of cretinism due to hormone deficiency. So many questions remain, such as where did their tools come from, and how did they, or did they, live alongside modern humans when these arrived 50,000–40,000 years ago?

There are also the enigmatic Denisova people. Since 2000, several human fossils including parts of a tooth, finger bone and toe bone have been recovered from Denisova Cave in the Altai Mountains, Central-East Asia. These remains are dated to around 40,000 years ago. There are also artefacts, including a bracelet, polished stones and pendants. Physical, chemical and genetic studies suggest the Denisovans may have descended from a wave of migration out of Africa independent of *Homo erectus* and modern humans, perhaps 1 million to 800,000 years ago. There is also evidence of common genetics between certain Neanderthals and the Denisovans, so perhaps a European group of the former evolved into the latter. It all goes to show the recent history of our human family probably had more extinctions than we realized.

Ways to the end

Today's humans are big-headed, both literally, and in the sense that some think we can adapt to anything. As discussed above, many other human species have gone extinct in the past. Why should we be so different, and could we cause our own downfall, or might something else spell the end for us? If we go extinct, could we leave newly evolved descendants?

A space-rock impact in the style of the end-Cretaceous mass extinction, perhaps coupled with vast tectonic activity and 'supervolcano' eruptions, could happen. Such events have recurred through Earth's history, but usually spaced apart by tens of millions of years. The chances of one within the next century or two are miniscule, although possible. Annihilation by nuclear weapons in a great war has been another 'favourite' for our termination. This threat has receded since the 'Cold War' era of the 1950s–70s.

● ONLY EXTANT HOMO
Only *Homo sapiens* survives from the many previous species of our genus. Will we go the same way as them?

Robins have increased in numbers as they thrive in urban and suburban settings. But songbirds less comfortable around human habitation are in drastic decline, some by 80 per cent in 30 years.

Some kind of global disease or pandemic is another suggestion. Modern medicine has a record of conquering diseases, for example smallpox by immunization, and massively fewer infections due to antibiotics. But microbe evolution means new infections continually appear, such as HIV/AIDS. The era of genetic engineering and modification brings new fears that some kind of 'superbug' might be created accidentally or on purpose, as part of medical research, and escape from the laboratory. It could then rampage around the world, spreading too fast and evolving too rapidly for us to defend ourselves. However, humans around the world are varied in their susceptibility to infection and their resistance to illness. Such a disease would be unlikely to affect all humankind.

Running out of resources

Over-population combined with dwindling natural resources is another scenario. Human population growth rate has slowed slightly since the 1950s–70s but our total number still broke through 7 billion in 2011–12. Projections show growth staying low or declining, yet reaching a total of between 8 and 10 billion by 2050.

When millions are malnourished and starving today, can the planet handle so many more people? It depends on how people handle the planet. Resources such as water, metal and mineral ores, and especially fossil fuel energy, are harder to locate and extract. Farmland is being degraded at alarming rates. However, more environmentally 'green' agriculture may gradually spread, and technological 'fixes' aim to switch from fossil fuels to more sustainable forms of energy – solar, wind, hydro-electric, geothermal, biomass, maybe waves and tides. It is a slow and painful process. Some experts predict that, as the situation gets more serious, our inventive and creative abilities will rise to the challenge.

Environmental crisis

The natural world provides us with many 'ecosystem services' – recycling of oxygen and air, water and nutrients, as described later (see pages 74-75). But our abuses may lead to a tipping point, perhaps involving climate change, global warming, sea-level rise, floods, droughts, habitat confusion, and more. Apart from the effects on wildlife, this could decimate croplands and the food chains that lead to ourselves. Again, some people have faith in our resourceful nature and our ability to get through such a predicament, even at the expense of many other species. Our sciences may progress to the level where we can interbreed or even genetically invent new animals and plants which fit our needs, and even influence our own future evolution.

● RAT AND COCKROACH
Brown rats *Rattus norvegicus* (left) and American cockroaches *Periplaneta americana* (right) may be undesirable but they are among the few species thriving on our leftovers and wastes.

If humankind continues to survive, even by lurching from one crisis to the next, it is always fascinating to imagine what our world may look like, say, in a few centuries. Sustainable energy, greenfield agriculture, enough clean water, and fairer ways of producing and sharing food may arrive. But many once-common songbirds are under threat now, so will there be a dawn chorus? Will the wildlife sharing our world be reduced to 'survivor' species who adapt to live so well alongside us – pigeons and starlings, rats and mice, foxes and raccoons, ants and cockroaches and houseflies, common weeds and similar tough plants? Not a very attractive vision, and future people may well look back and wonder why the generations of today let it slide so far.

why bother?

THE IMPORTANCE OF EXTINCTION TODAY

● TIGER
Millions are spent annually on tiger conservation. But it's not enough, with millions more supporting the illegal trade in its body products.

SHOULD WE SAVE THE TIGER? Around much of the world, people would answer resoundingly 'Yes!' The public's will, especially in rich countries, is right behind efforts to save this magnificent predator. We admire its size, strength, beauty, grace, cunning and terrific killing power. Conservation efforts for the tiger are huge. The Convention on International Trade in Endangered Species of Wild Fauna and Flora (CITES) bans international trade in tigers, their body parts and any derivates, without special licence (such as for scientific research). The 13 nations where tigers still occur naturally in the wild are signed up to CITES and are multiplying their efforts to clamp down on poaching and trade within, as well as between, their national boundaries.

Organizations fighting to conserve tigers include Project Tiger, founded in 1972, which has helped to establish more than 25 reserves in India. Save the Tiger Fund, set up in 1995, has contributed more than US$ 10 million to efforts. The Global Tiger Initiative, WWF, Tigers Forever (a collaboration between the Wildlife Conservation Society and Panthera), and many other organizations are heavily involved in tiger conservation.

But the striped big cat is in deep trouble. WWF estimates its numbers have fallen by 90 or even 95 per cent in the past century. As described earlier (see page 7), three of its subspecies have gone extinct. Global numbers of all remaining subspecies of wild tigers are estimated at an all-time low of 3,000–4,000. Their habitats are dwindling and also being fragmented. As a top or apex predator, an adult male tiger needs a hunting range of as much as 100 square kilometres (just under 40 square miles), to catch sufficient prey. Such large continuous areas of suitable habitat are disappearing fast. Setting up 'tiger corridors' to link reserves and other suitable areas may help tigers to disperse more widely and prevent the many problems of inbreeding.

Downward trend

Despite legal protection, tigers are still killed, especially in remote rural areas. Because of ever-growing human populations, and expansion of agriculture, wild space is shrinking. Some tigers come into close proximity with villages, where people take action against them in fear of attacks on their livestock or themselves. Also the illegal but highly lucrative trade in tiger skins, bones, teeth and other parts can bring huge profits, tempting tiger poachers to satisfy the supply chain. Clamping down on the trade, both national and international, is a top priority for tiger conservation. Otherwise worst-case scenarios predict that tigers, if not extinct, could fall below viable population numbers for the species to survive within 50 years, maybe 25, or even less.

Because of this big cat's massive fame and worldwide appeal, failing the tiger would send shock waves around the world. It is seen as a symbol of humankind's inability to look after the world in our role of 'guardians of the planet'.

Some of the world's earliest artworks include Australian Aboriginal rock paintings celebrating animals such as kangaroos, dingoes and emus.

The right thing to do

Herein lie some of the important moral and ethical reasons why conservation is important. Protecting biodiversity is the 'right' thing to do. It conveys a sense of care and duty so that others can enjoy the natural world, both today and for future generations. Most parents strive to do what is best for their offspring, and overall this includes bequeathing a healthy planet, with the awe and beauty of amazing wildlife It is part of the continuing tradition of natural heritage, passing on a rich and diverse world. Mingled with these thoughts are the breathtaking beauty, dazzling variety and sheer wonder of the natural world – aesthetic glory for average citizens and for creative geniuses. Through the ages artists, sculptors, writers, musicians and actors have been inspired by nature to superlative works.

Yet saving wildlife and preventing extinction should not be for a few high-profile, headline-grabbing species such as big cats, great apes, huge whales, and others with maximum aesthetic appeal. There are also vital biological and ecological reasons. A diverse, balanced ecosystem is stronger, more resilient to disturbance, and more adaptable to change. As described through many examples in this book, extinctions can disrupt food chains and the balance of nature, and lead to ecosystem collapse with knock-on effects that might impact greatly on ourselves.

● ARTISTIC INSPIRATION 2
Albrecht Dürer's *Wild Hare* (1502) is a famously stunning example of semi-photographic realism from nature.

● NENE
Still a major cause of death for the nene in Hawaii is road kill by speeding motorists.

The nene or Hawaiian goose, *Branta sandvicensis*, is, or was, the world's rarest goose. By the late 1940s there were fewer than 50 left in the Hawaiian islands, from an original population of more than 20,000. The usual island-introduced species such as cats, and also especially in this case the mongoose, as well as hunting by people, had wrecked its populations. In 1957, to raise the profile of its plight, the nene was adopted as the national bird of Hawaii.

From 1950 skilled conservation work, in Hawaii and at the Wildfowl and Wetlands Trust in Slimbridge, southwest England, managed to encourage the nene to breed in captivity. Great efforts were made to restore suitable habitats in Hawaii. Releases started in 1960. The work was so successful that in 2011 the release of captive-bred nene could end. Today more than 2,000 of these attractive geese again grace Hawaii, with further populations in parks and reserves around the world. However, there is still concern that, coming back from such a small population, there could be a limited amount of genetic variety within the species – yet another area scientists are currently investigating. In 2012 the IUCN assessed the nene as 'Vulnerable'.

Traditions and cultures

Many animals and plants threatened with extinction are an integral part of a country's or community's identity. They hold a revered position in centuries-old traditional events, celebrations and worship. In modern times, famous mammals, birds and plants have been adopted by states and nations – but then they drifted towards rarity and threat.

● **AMERICAN BISON**
Yellowstone National Park is now a secure home to some 3,000 American bison.

● **ARFAJ**
This thorny-leaved shrub can thrive in arid habitats if grazed, provided the disturbance is well controlled.

In such cases, conservation issues are even more sensitive. The giant sable antelope of Angola, mentioned earlier (see page 8), and the nene are examples.

The arfaj, *Rhanterium epapposum*, a low shrub with striking yellow blooms, is the national flower of Kuwait. It was selected both for its beauty and its usefulness, as a source of natural remedies, bedding and firewood. But the arfaj and much of its semi-desert habitat are greatly threatened by overgrazing. Nomadic livestock like the arfaj's taste in particular and soon overgraze it. Conservation work has shown that, if grazing is limited or controlled to certain times, the arfaj and other plants soon recover. Grazing becomes available again and a balance results.

Being an important part of tradition and folklore has also had the opposite effect – hastening species to near-extinction. The American bison, *Bison bison*, was a keystone and engineer (see page 48) species on the vast prairie grasslands. Its dramatic decline in North America during the nineteenth century was mainly due to mass slaughter by rifle for skins, meat and 'the fun of it', helped by habitat loss as prairies were ploughed and forests chopped down. At one time bison-hunting was encouraged by the government to 'disadvantage' some groups of Plains Native Americans, who lived alongside bison, used most of its body parts, and incorporated the animal into their folklore.

From millions of bison, numbers were fewer than a thousand by 1905. The first conservation efforts were private, as concerned individuals established small herds on their large ranches to avoid total extinction. The US Government slowly realized the situation and included bison in its plans for a network of national parks, and state authorities did the same. Today bison are safe, but their health and breeding potential may be affected by the 'genetic bottleneck' of so few survivors, coupled with interbreeding and 'genetic pollution' from domestic cattle.

Legend of the Thunderbird

Another famous 'back-from-the-brink' species is the majestic California condor, *Gymnogyps californianus*. It is a huge scavenger–predator with a wingspan of 3 metres (10 feet). But in its homeland of southeast North America, its numbers fell drastically. Causes included lead poisoning from scavenging carcasses of creatures containing lead shot, being shot itself in case it took livestock, power-line collisions, and in some cases from the 1950s, DDT poisoning. The infamous insecticide DDT accumulated through food chains from bugs to top consumers like the condor, where it caused thinned eggshells. These broke easily in the nest, killing the immature chick within.

This largest of North American birds held great importance to many Native American peoples. It may be one inspiration of the 'Thunderbird' myth – a huge bird of legendary strength and power, whose wingbeats cause thunder. Some Native groups regard the condor as sacred, using its feathers in ceremonies, and telling age-old tales of its exploits. Finally the US authorities made a bold decision to save the species. In 1987 the entire population of only 22 condors was trapped for captive breeding, leaving none flying free. The plan was controversial but it worked. From 1991 condors were released into the wild. By 2012 their numbers had risen to about 225 wild and 180 captive.

● CALIFORNIA CONDOR
Post-mortem studies of condor carcasses to determine cause of death help to direct its conservation programme.

SPOTLIGHT ON... WILD CABBAGE AND WILD APPLE

Vegetables such as broccoli, cabbage, cauliflower, Brussels sprouts and kale are all cultivated varieties, or cultivars, of the wild cabbage, *Brassica oleracea*. Likewise the wild apple tree, *Malus sieversii*, has given us many varieties of apples to suit almost every taste, as well as apple derivates like cider.

Wild apple trees are now being examined again for high-tech interbreeding with other apples, to produce new, better varieties – for example to withstand more extreme conditions such as cold and drought. The wild cabbage, originally from southern and western Europe, is now found in Britain, France, Spain and Germany. It can tolerate salty and lime-rich conditions, and again research programmes are looking at using it to cultivate newer, tougher varieties, such as for better salt and disease resistance. No one knows what new and beneficial features may emerge from the gene sets of these wild ancestors. Yet both plants are under threat in nature. The IUCN lists wild cabbage as

● WILD ANCESTORS
Wild cabbage (above left) and wild apple are among several crop ancestors re-examined for useful genes.

'Data Deficient' with populations decreasing and in need of action, and wild apple as 'Vulnerable' in its native Central Asia.

● MUSK DEER
Male musk deer are cruelly maimed or simply slaughtered for their musk, used in the perfume industry.

Ecosystem services and products

Natural ecosystems and the species they support provide 'free' services for humankind – from food to natural medicines and crop pollination. If we reduce biodiversity by human-induced extinctions, we lessen nature's ability to continue these benefits, which are usually described in four categories:

- **Provision** – foods from the wild such as seafoods and game, also crops bred and livestock domesticated from wild ancestors; fresh clean water; raw materials such as minerals for industry; remedies and medical products; timber for building and fuel; fibres of many kinds, from silk and wool to cotton and flax; and energy such as biomass fuels and hydro-electricity.
- **Regulation** – absorbing carbon from the atmosphere by plants; decomposing wastes; refreshing the atmosphere with oxygen by plants; natural control of pests and diseases; and crop pollination.
- **Support** – recycling and spreading nutrients; dispersing seeds; and plant growth itself, known as primary production.
- **Cultural services** – inspiration; recreation; and scientific research and discovery.

The concept of ecosystem services lies behind the founding of the IPBES, mentioned earlier (see page 39). The stark reality is the more we abuse nature, and threaten extinctions, the more we risk losing the essential provisions and services on which we rely for survival.

Exotic products

As well as basic foodstuffs, textile fibres and fuels, these natural services include the rare and exotic. For instance, herbs and spices are not vital for human life, but they enrich our eating experiences so much that they provide the 'spice of life'.

The substance musk has been treasured through history in perfumes, aphrodisiacs, incense and medicinal remedies. It is produced by male musk deer in a gland towards the rear of the body, as part of their mating process. Traditionally it is gathered by catching and killing the deer and removing the gland or 'pod'. Careful removal of the gland to leave the deer alive is possible, but collectors seldom take the time or trouble. As a consequence all seven species of musk deer are widely hunted. The Siberian musk deer, *Moschus moschiferus*, is classed by the IUCN as 'Vulnerable', and the other six species are all 'Endangered'.

Another luxury wild product is the shahtoosh shawl, woven from an incredibly fine, soft wool derived from the underfur of the Tibetan antelope, *Pantholops hodgsonii*. Traditionally local people hunted the antelope for this wool and also meat, bones for tools and utensils, and other products. It is possible to harvest the wool from live animals. But again modern hunters, working quickly for profit, find it easier simply to kill the deer, take the wool, and waste the carcass. The antelope's population has plummeted by 90 per cent in the past century, and by 50 per cent just from 1990. The IUCN lists it as 'Endangered'.

● **TIBETAN ANTELOPE**
As well as fur, Tibetan antelopes are poached for their horns, ground to make traditional but ineffective 'remedies'.

● **BEE POLLINATION**
A common carder bee, *Bombus pascuorum*, visits a plant – and provides the service of pollination which is vital to its whole ecosystem's well-being.

● ORANG UTAN
Despite protection, orang utans are trapped as exotic pets or for body parts used in customs and 'medicines'.

Better alive than dead

In Victorian times, many rare species were pushed to final extinction by the scramble to grab the last few specimens as sought-after prizes by collectors, museums and trophy-hunters. Beautiful butterflies, fascinating frogs and colourful birds – complete with nests and eggs – met their end like this.

Attitudes have changed and now rare or spectacular creatures are worth more as ecotourist attractions. In its best form, ecotourism has very little impact on target animals, plants and general wildlife in its location. It also raises funds and brings resources for continued conservation work, benefits local people and habitats, educates visitors and encourages all-round responsible, positive attitudes to wildlife.

Whale-watching, safaris and ocean cruises are popular with ecotourists. Specialized ventures look at our close living relatives – chimps, gorillas and orang utans. Tours to see Bornean orangs, *Pongo pygmaeus*, and the Sumatran species *Pongo abelii* – IUCN-designated as 'Endangered' and 'Critically Endangered' respectively – are a growth industry. But as with any ecotourist adventure, there is a danger of more becoming less, as rising visitor numbers pressurize local resources, disturbing the species and ecosystems they once aimed to preserve.

The mountain gorilla, *Gorilla beringei beringei*, is a subspecies of the eastern gorilla. Fewer than 500 survive at high altitudes in Central Africa. Visits are highly supervised, far from cheap and subject to interruption in this volatile region. But the money they bring in is a vital part of saving the mountain gorilla and its habitat. People who complete the mountain gorilla voyage are rewarded by seeing a creature that, 50 years ago, was by now expected to have gone extinct.

● MOUNTAIN GORILLA
After decades at the threshold of extinction, mountain gorilla numbers are increasing. But it is a slow and fragile recovery, and hugely dependent on the modern eco-tourism trend.

back from the brink
EXTINCTION AND CONSERVATION

IN THE FACE OF TODAY'S PLETHORA of threats and impending extinctions, there are plenty of success stories to warm the heart. Conservation initiatives and green agendas give cause for hope. As catalogued on many pages in this book, species have been, and are being, pulled back from the precipice. But it takes time, resources, goodwill, cooperation, local involvement, motivation – and money. All of these are limited. How do we select which species to save, and which methods are best?

● GOLDEN LION TAMARIN
More than 150 zoos, sanctuaries and captive breeding centres worldwide have helped this endearing monkey.

Enter the concept of the flagship species. This is usually an animal — often a mammal, perhaps a bird, rarely a reptile, fish or insect. It tends to be big, which means it needs large areas of habitat to survive. It has mass appeal and a high profile with the public, due to features such as ferocity, grace, cuddliness, and so on. As well as being a biological species, and a flagship species, it may also be a keystone species, and perhaps an indicator species (see page 48).

The flagship species is chosen to headline a conservation campaign to head off its extinction. Saving it in the wild means preserving where it lives. In this way great areas of habitat become the real target, and within them are all the co-existing species of animals and plants. Tigers, lions, bears, giant pandas, elephants, rhinos, apes, monkeys, eagles, crocodiles and sharks are all examples. If the campaign is successful the whole ecosystem benefits, in the way that an initiative headed by a species of slug or fern, while perfectly valid, could never achieve.

Save the tamarin

The golden lion tamarin, *Leontopithecus rosalia*, is a small, attractive, elaborately maned monkey. It inhabits tropical rainforests along the Atlantic coast of Rio de Janeiro State, southeast Brazil. After four centuries of logging, mining, polluting and urbanizing, its habitat was irreversibly ruined and fragmented. The monkeys themselves were also captured as exotic 'handbag' pets. In the late 1960s the species' charisma, petiteness and vulnerability brought its plight to world attention. In that decade there were probably fewer than 500 golden lion tamarins left. Like the giant panda and tiger, it became an icon of conservation. Some were captured for breeding in zoos and wildlife parks, while a massive campaign tried to save the tiny remaining fragments of its habitat.

The tamarins are now slightly safer. Captive breeding and reintroduction have led to three wild populations totalling more than 1,000, with another 500 or so in zoos and specialist sanctuaries. In the process, this flagship species has helped to save many thousands of hectares of coastal rainforests, with their millions of other inhabitants. But threats still loom, especially spreading urbanization. Scientists estimate that even if all the remaining suitable habitat was occupied by the tamarins, their total population would still not reach 2,000. This is known as the carrying capacity of an environment — how many of a particular species or group can be supported in a certain area, in terms of food, living space, hiding places, breeding sites and the other resources it needs.

The limited carrying capacity of the residual Atlantic forests means that tamarins might never become numerous enough to maintain a healthy variety of genes and avoid the perils of inbreeding. This is why its captive breeding remains so important. In 2003 the IUCN eased the golden lion tamarin's assessment from 'Critically Endangered' to 'Endangered', but the cute flame-haired monkey stays highly threatened and dependent on continuing conservation.

In 1972 the antelope called the Arabian oryx, *Oryx leucoryx*, was believed extinct in the wild, partly as a result of hunting for its body parts and impressive 70-centimetre (28-inch) horns, and also because of live capture for private collectors. The national animal symbol for Jordan and Qatar, it has a strikingly beautiful white coat, stands 1 metre (3 ⅓ feet) tall at the shoulder, and weighs around 70 kilogrammes (155 pounds).

The oryx's plight had already been recognized. In the early 1960s nine oryx – some captured wild, some from private collections and other zoos – began breeding at Phoenix Zoo, Arizona, USA. Operation Oryx was one of the first captive breeding programmes for any animal. Phoenix was a chance location, since the originally chosen breeding site was in Kenya, but a disease outbreak meant a change of plan.

Success meant that reintroductions began in 1982 in Oman, spreading to several other nations across the Middle East. There are now probably more than 1,000 Arabian oryx in the wild across a dozen locations, with more than 5,000 in parks and reserves. They have full legal protection and the highest CITES status. In 2011 the IUCN downgraded the oryx's status from 'Endangered' to 'Vulnerable'.

● ORYX
Oryx conservation began in 1962. It was 'Extinct in the Wild' in 1972 and reintroductions date from 1982.

Four to more

The Mauritius kestrel, *Falco punctatus*, is a fleck-chested bird of prey with a wingspan of 45 centimetres (18 inches). In 1974 only four of these birds were known. The causes of such catastrophic decline are depressingly familiar, especially for islands like Mauritius – invasive animal species such as the mongoose, cat, black rat and crab-eating macaque monkey; along with degradation of 95 per cent of suitable habitat, partly as a result of invasive plants like Chinese guava; and eggshell thinning due to the pesticide DDT, used to tackle malaria epidemics among humans.

● MAURITIUS KESTREL
This petite raptor hunts large bugs like crickets and cockroaches, also little lizards and small birds. The last release of captive-bred birds was in 1994.

Energetic conservation measures from the late 1970s included egg removal from nests – done properly, the birds lay again – for incubation and hatching. Other measures included extra food, nest boxes and varied nest sites, and predator control at nesting and release areas. Nests were monitored and, when the parent birds seemed to be having trouble, eggs and chicks were taken for rearing, using common kestrels as foster parents. Several parts of the kestrel's range are now conservation areas, and planned roads and other developments here have been cancelled. These amazing efforts have brought current populations of the Mauritius kestrel to more than 400, although these fluctuate and sometimes fall. There is potential for numbers to rise to the carrying capacity for what remains of its rocky forest and cliff habitats, that is, probably 900–1,000 birds.

In situ conservation is helping and saving wildlife in its natural habitat, which means preserving the habitat itself. *Ex situ* conservation involves helping a species away from its natural habitat. The latter is usually a last resort since it is so difficult to re-create all of the original conditions needed for a big cat or eagle.

Combined conservation

The Lord Howe stick insect, *Dryococelus australis*, is a large, stout, heavy-bodied example of its group, up to 15 centimetres (6 inches) long – and a fine example of a 'back-from-the-dead' Lazarus species. Its strong body casing have led to the nickname

● LORD HOWE STICK INSECT

The adult Lord Howe stick insect, although wingless, is powerful, protected and resilient. It was once common enough to be used by local fishing crews as bait.

'land lobster'. It was once common on Lord Howe Island, 600 kilometres (about 375 miles) east of Australia. A shipwreck here in 1918 probably allowed black rats to infest the island and devastate these flightless insects, along with many other small animals. By 1930 the Lord Howe stick insect was believed extinct.

In 1964 rock climbers visited a rat-free sea-stack known as Ball's Pyramid, about 20 kilometres (12 miles) from Lord Howe Island – and found a large stick insect. Unfortunately it was dead, but it raised hope of a resurrection. After several further visits, scientists were amazed to locate approximately 24 live Lord Howe stick insects subsisting around a single bush. Two pairs were collected in 2003 for *ex situ* captive breeding. Melbourne Zoo successfully reared the young, first a few dozen, then many hundreds, and now thousands. In 2012 the zoo offered 20 schools the chance to breed these celebrated stick insects in their classrooms. Full release back onto Lord Howe Island, for further *in situ* conservation, will wait until the black rats are gone, although test releases into a few specially prepared areas are encouraging. The IUCN revised this species from 'Extinct' to 'Critically Endangered'.

EDGE species

Another reason to focus on a particular species for conservation is the criterion known as EDGE – evolutionary distinct and globally endangered. The EDGE of Existence programme has been developed by the Zoological Society of London. It raises awareness and encourages conservation of species that are not only on the verge of extinction or highly threatened, but also highly unusual, with few close relatives and great distinction in their anatomy, behaviour and other features. Each EDGE species represents a unique evolutionary history, different from other living things, which is in great danger of being lost for ever.

EDGE teams use particular criteria to score the conservation value of individual species. As well as the IUCN threat category, there are factors such as how long the species has been evolving along its own particular pathway, and how few other, closely-related species still survive. For these reasons, so-called 'living fossils' score highly. So far the EDGE methods have been applied in detail to mammals and amphibians.

One of the top EDGE amphibians is Archey's frog, *Leiopelma archeyi*, from a small area in the northeast of North Island, New Zealand. It is uncannily similar to frogs known only from fossils 150 million years old. Archey's frogs live wholly on land. After the female lays her eggs in a damp place, the male guards them until they hatch into froglets with tails. These ride him piggyback-style for several weeks as they continue to develop.

Fungal pandemic

Archey's frog populations have completely disappeared from some areas, and have been reduced by 90 per cent in others. The species is threatened by introductions such as rats, probably climate change – and, as with many other amphibians worldwide, by the disease called chytridiomycosis. This is infection by the fungus or mould *Batrachochytrium dendrobatidis*, known as Bd for short, from the chytrid group of fungi, which develop their reproductive spores in pot- or tube-like capsules. The Bd fungus is hardy, lives in water and soils – but also grows in the upper surface layers of amphibian

● ARCHEY'S FROG
Plans to mine areas of Archey's frog habitat have been dropped, but captive breeding proves difficult.

● **GOLDEN TOAD**
This toad, here a male, was among 20 amphibian species to disappear in 50 years from a specialized high-altitude cloud forest habitat near Monteverde, Costa Rica.

skin, leading to thickening, swelling, sores and ulcers. Amphibians 'breathe' through parts of their skin and also absorb water and important natural salts and minerals. Chytridiomycosis interferes with these skin functions and leads to odd behaviour in the infected creature, such as lethargy even in the face of imminent danger.

Chytrid fungus has already helped to kill off many frogs and toads. One was the golden toad, *Incilius periglenes* (formerly *Bufo periglenes*), of Costa Rica, Central America. Just 5 centimetres (2 inches) long, the males of this species were a bright golden-yellow, while the females were black or dark green with bright red patches. Golden toads were only discovered and named in 1966, and had a tiny range of less than 10 square kilometres (4 square miles) in upland forests near the town of Monteverde. This species was probably affected by climate change and air pollution as well as by the Bd fungus. The last sighting was in 1989 and the toad is now regarded by the IUCN as 'Extinct'.

The chytrid or Bd fungus possibly spread from its origins in Africa through world trade in the clawed frog/toad, *Xenopus laevis*, a common laboratory animal in scientific research and medical tests, including human pregancy tests. It might then have been spread further by the international trade in frogs for food and as exotic pets. In New Zealand, the Bd fungus was first detected in the wild in Archey's frogs in 2001. Since then the overall fall in this frog's population is estimated at 80 per cent. A multi-pronged conservation effort includes research into captive breeding, saving its habitat, and removing or controlling predators and competitors. Scientists are also taking part in the massive global campaign to find treatments or cures against the Bd chytrid fungus. This fungal threat to amphibians was only identified in 1998 and is now thought to affect up to one-third of amphibians, and be a major factor in the extinction of more than 120 species.

Unique deer

Another very distinctive species is Père David's deer, *Elaphurus davidianus*, the only living member of its genus. It is named after the French missionary Armand David. While working in China he recognized the deer's uniqueness, and in 1866 sent carcasses back to Paris for description and naming. After centuries of heavy hunting, only one herd of the deer then existed, in the Chinese Emperor's vast hunting garden near Beijing. By the early 1900s a series of catastrophes, including a flood, and eating of deer by peasants and then soldiers, wiped out that herd.

Luckily some Père David deer had been transported to European zoos and collections. The 11th Duke of Bedford, Herbrand Russell, gathered as many as he could and established a herd at Woburn Abbey, in southeast England. They bred well. This saved the species and allowed deer to be exported to other zoos and parks, including back to the original hunting garden, reserves and paddocks in China. Future plans may allow this deer at last to be released into the wild.

Success and failure

Wildlife conservation, and especially saving species from extinction, is an immense and complex business. It needs huge organizing skills, experts on board, surveys and field studies, perhaps captive breeding set-ups and habitat protection, all financed in many ways, from government grants to corporate sponsorship and public fundraising. And it takes a long time.

● PERE DAVID'S DEER
Milu Park near Beijing – 'milu' is one of the species' Chinese names – is a relocation site for Père David's deer.

● FLOREANA MOCKINGBIRD
Floreana mockingbirds collected during Charles Darwin's visit to the Galapagos islands on HMS *Beagle*, are used in research to understand the genetic diversity of the current population.

In 2007 a 10-year campaign began to save the Floreana mockingbird, *Mimus trifasciatus*. It disappeared from Floreana, an island in the Galapagos, in the eastern Pacific, by the 1880s – the usual culprits for predation and habitat destruction being rats, cats, dogs and goats. However, these plagues never reached two small nearby islets, and the mockingbirds clung on here.

This bird is especially valued since it was studied by naturalist Charles Darwin when he visited the Galapagos in 1835, during his round-the-world voyage on board HMS *Beagle*. He noticed that the mockingbirds on Floreana were similar, but noticeably different, to those on other islands and back on the South American mainland. It was these birds, rather than the more famous Galapagos finches, that triggered his thinking about the process of evolution by the mechanism of natural selection. It led to one of the most important books in science, *On the Origin of Species by Means of Natural Selection* (1859).

Original mockingbird specimens from the *Beagle*'s voyage, held by London's Natural History Museum, are part of ongoing research into the genetics of the current population. They have helped conservationists to understand the genetic diversity of the modern birds by comparing them with the historic mainland population. This research contributes to the species' decade-long rescue plan, masterminded by the the Charles Darwin Foundation. It involves more than 20 complex action points, such as guarding the remaining mockingbirds, and changing conditions on Floreana by eradicating or controlling rodents, cats, goats, pigs and donkeys, ready for the bird's reintroduction. There are fewer than 100 Floreana mockingbirds left, and they are IUCN-assessed as 'Critically Endangered'.

Intercontinental conservation

Conservation efforts extend across great oceans and wide continents. London is a long way from Mexico. But the Aquarium at ZSL (Zoological Society of London) London Zoo is spearheading a campaign to save four Mexican species of fish. Pupfish live in small pools among rocks, many in the middle of deserts and dry scrubland. These pools are small and isolated, each with its own conditions such as water temperature and salinity, and its own unique species of fish and other wildlife – truly an ecosystem in miniature. These tiny locations are under threat from water drained for agriculture and human use, and introduced species. ZSL has re-created sets of conditions for four species – the chequered pupfish, *Cualac tesselatus* (IUCN 'Endangered'), Charco Palma pupfish, *Cyprinodon veronicae* ('Critically Endangered'), and Potosi pupfish, *Cyprinodon alvarezi*, and La Palma pupfish, *Cyprinodon longidorsalis* (both 'Extinct in the Wild').

Another international effort concerns the spoon-billed sandpiper, *Eurynorhynchus pygmeus*, an attractive small wader of East Asia. Always staying on or within a few kilometres of the coast, it breeds in the Siberian north and migrates to South and Southeast Asia, from India to the Philippines and Malaysia. At least it did. Many of its stopover wetlands en route have been 'reclaimed', that is, drained for agriculture, industry and human settlements, and/or polluted. In its winter areas trapping is a huge problem. The bird's decline has steepened in the past few decades, falling by more than 80 per cent between 2002 and 2012. Probably fewer than 400 spoon-billed sandpipers are left. In 2011, 13 of the birds arrived in specially designed quarters at the Wildfowl and Wetlands Trust in Slimbridge, southwest England – mentioned above (see page 71) in connection with the nene. Also a batch of spoon-billed sandpiper eggs was brought here in 2012 and chicks hatched in captivity at Slimbridge. They represent the main hope for this IUCN-listed 'Critically Endangered' species.

● **SPOON-BILLED SANDPIPER**
The wide-tipped or spatulate beak of this sandpiper, seen in this valuable reference specimen, works as a mini-shovel when feeding.

● **PUPFISH**
Like its Mexican counterparts, the desert pupfish *Cyprinodon macularius* is very localized and highly threatened.

The kakapo

Another bird conservation effort that has achieved an amazingly high profile is the campaign to save the kakapo, *Strigops habroptila*, of New Zealand. It is an especially remarkable bird – the heaviest parrot, the only flightless parrot, nocturnal, with fibrous feathers almost like whiskers around its face. With no ground-dwelling mammals in New Zealand, as mentioned previously (see page 58), the kakapo evolved into the niche of a night-active herbivore and became common in many habitats. First the Maori people with dogs and rats, then Europeans with cats, stoats and other introduced predators, plus huge land-clearance programmes that involved setting fire to vast areas, crippled its numbers. As its rarity increased, the bird's unusual appearance and quaint features made it a favourite with collectors, museums and zoos, who rushed to gather specimens.

Conservation measures began in the 1950s but were haphazard, uncoordinated, and hampered by lack of knowledge about the bird's behaviour and ecology. By the early 1970s the kakapo was suspected as extinct or near enough. Then in 1977 it was reported on Stewart Island (Rakiura), off the southern tip of New Zealand's South Island. The Kakapo Recovery Plan began in 1989 with transfer of kakapo to predator-free reserve islands. These are now monitored for appearance of predators. Also, research showed that female kakapo breed only every few years to synchronize with 'mast' years when local rimu trees produce extra-plentiful seeds, which the birds eat. Feeding kakapo specially chosen substitute foods such as almonds, walnuts and brazil nuts encouraged breeding more often.

Today there are more than 120 kakapo. Each has a name and a radio-tag transmitter. The birds' lives and activities are followed daily by their legions of human fans. Selected islands are being cleared of introduced predators so that the kakapo – assessed as 'Critically Endangered' by the IUCN – and other native species can live in safety.

● **KAKAPO**
This highly unusual, heavy-bodied parrot reaches 60 cm (2 ft) in length and weighs up to 4 kg (9 lb).

SPOTLIGHT ON... BLUEFIN TUNA

The Atlantic bluefin tuna, *Thunnus thynnus*, can exceed 4 metres (13 feet) in length and 500 kilogrammes (1,100 pounds) in weight. Fast and powerful, it hunts small schooling fish such as anchovy, saury (mackerel pike) and hake. It is one of the few 'warm-blooded' fish. The bluefin can generate body warmth so that its muscles work more efficiently, allowing it to race along at more than 60 kilometres per hour (37 miles per hour).

But giant fishing fleets are driving these great fish and many others ever downwards. Tuna is a staple food in many countries, and demand grows because of its perception as a healthy option. Estimated global tuna decline is as much as 50 per cent in the past 40 years. In some areas Atlantic

● BLUEFIN SHOAL
Bluefin tuna stocks depend on their prey species, which are also being rapidly fished away.

bluefin numbers are down more than 80 per cent in this time. Its IUCN ranking is 'Endangered'.

Conservation campaigners lobby for reduced tuna quotas, but the authorities in charge of such matters act slowly, if at all, under great pressure from the fishing industry. In any case, fishing far out in the ocean is notoriously difficult to monitor, and annual quotas are routinely exceeded. Making consumers aware of the perils facing this and other tuna species, plus expansion of tuna farming in sea-pens, are among its hopes.

Clams and conch

All kinds of creatures and plants are subjects of conservation – even relations of common snails. Giant clams, *Tridacna gigas*, from Southeast Asia and the West Pacific are famed for their size, with shells measuring more than 1 metre (3 ⅓ feet), their weights, in excess of 200 kilogrammes (440 pounds), and their longevity – over 100 years. They have become rare in the wild due to collection for food and trophies. However, scientists have devised ways of breeding the clams in captivity, for food and also to release into the wild as stock replenishment.

Another shellfish, the queen conch, *Lobatus gigas*, from the Caribbean, is suffering the same fate. Its elaborate shell is especially prized by collectors. But harvests have long been unsustainable. The Caribbean's International Queen Conch Initiative aims to raise awareness of this fascinating creature, reduce overfishing and improve the prospects for its captive breeding and farming.

● GIANT CLAM SHELL
Because of their scarcity, scientists have developed ways of raising these shellfish in clam farms, for sale as meat and for release into the wild.

Back from the dead?

Is extinction truly for ever? Modern techniques in genetics, selective breeding and surrogate motherhood may raise hopes that an extinct species could literally be brought back to life. The Pyrenean ibex, *Capra pyrenaica pyrenaica*, is a subspecies of the Spanish ibex. The last survivor, Celia, died in 2000. However, samples taken from her have been used in cloning experiments to produce embryos, which are then implanted into surrogate mother goats. In 2009 one of these pregnancies

● PYRENEAN IBEX
An attempt to resurrect this subspecies by cloning resulted in a live birth but the young died within minutes.

● QUAGGA AND FOAL
Genetic studies in 1980 showed the extinct quagga could perhaps be 'bred back' from living zebras. Quagga foals developed their limited stripes only after several months.

proceeded to term and the kid was born alive. But it survived for only seven or eight minutes due to lung problems.

The quagga, *Equus quagga quagga*, from South Africa was another extinct subspecies, this time of the plains zebra. It had stripes only on its front half, fading away to the plain brown rear quarters. Quaggas were hunted out of existence in the wild by the late nineteenth century, and all zoo specimens had died by 1883. In 1987 the Quagga Project began. Its aim is to select the most quagga-like plains zebras available for reproduction, with the hope of 'breeding back' the distinctive appearance. Results are gradual but by 2012 the herds numbered more than 100 across 11 locations and the offspring are coming closer to the striped>stripeless look. Much discussion centres on whether these animals are a true re-creation of the actual quagga, or merely look-alikes.

More fanciful resurrections involve, for example, cloning a mammoth from long-frozen carcass tissues and using elephants as surrogate mothers. Or even a Neanderthal with a modern human mother. The gene sets or genomes of these species are available. The costs, resources and technical difficulties are colossal. Yet so are the advances made since Dolly the sheep was born in 1996, as the first mammal cloned from a body cell (rather than egg or sperm).

● WOOLLY MAMMOTH
This massive herbivore from the most recent Ice Ages fell by the wayside. Probably it could not cope with climate change and hunting pressure from a newer species – ourselves. Today the former is again headline news.

Dangerous times

In 2012 the IUCN Red List of Threatened Species underwent its regular update. It contains more than 63,500 species assessed, with about 20,000 considered threatened or on the brink of extinction – including one-third of conifer plants, the same proportion of reef-building corals, two-fifths of amphibian species, one in seven birds, and one in four mammals. Biodiversity and ecosystem health are at risk as never before.

Throughout Earth's immense prehistory, extinction has been the end for some, yet allowed a beginning for others. Mass loss of species has been followed, after an interval of a few million years, by a return of biodiversity. But current extinctions are happening rapidly. Rich nations show concern and encourage action, however in poorer regions pollution, poaching and similar dangers are rife. In some places habitat loss is out of control, fuelled by the consumer demands of the wealthy, and aided by greed and corruption. Natural resources like petroleum, coal and mineral ores are diminishing fast. Climate change looms large for everyone. The ecosystem services that nature provides for us may at some point collapse. This could hasten our fate as one of the millions of species that came and went, although in this case, we could also take a sizeable number of other species with us – sooner rather than later.

Learning lessons from the past, it seems very unlikely that the human species can cheat its eventual extinction. But we can avoid a premature end-of-the-world scenario. With massive and concerted action now, we could protect much of Earth's biodiversity, as well as rescheduling our own demise – later rather than sooner.

index

picture credits